95 Advances in Polymer Science

Polymer Physics

With contributions by
H. Biederman, G. Kothe, M. Lazár,
A. Ya. Malkin, K. Müller, Y. Osada, R. Rado,
J. Rychlý, K.-H. Wassmer, P. V. Zhirkov

With 83 Figures and 14 Tables

Springer-Verlag Berlin Heidelberg GmbH

ISBN 978-3-662-15039-9   ISBN 978-3-540-46909-4 (eBook)
DOI 10.1007/978-3-540-46909-4

Library of Congress Catalog Card Number 61-642

© Springer-Verlag Berlin Heidelberg 1990
Originally published by Springer-Verlag Berlin Heidelberg New York in 1990
Softcover reprint of the hardcover 1st edition 1990

2152/3020-543210 — Printed on acid-free paper

Editors

Table of Contents

Dynamic Magnetic Resonance of Liquid Crystal Polymers: Molecular Organization and Macroscopic Properties

K. Müller, K.-H. Wassmer* and G. Kothe
Institut für Physikalische Chemie, Universität Stuttgart, Pfaffenwaldring 55,
D-7000 Stuttgart 80, FRG

Thermotropic liquid crystal main and side chain polymers (LCPs) have been studied by dynamic magnetic resonance techniques, including continuous wave ESR and pulsed deuteron NMR. Analysis of the various experiments, employing an appropriate relaxation model, provides detailed information about the dynamic organization of these systems. The discussion reveals the prominent molecular features distinguishing LCPs from their monomeric analogues. Concomitant investigations of the bulk behaviour indicate that the unique macroscopic properties of these polymers are strongly correlated to the specific molecular properties, determined by magnetic resonance techniques. In summary, the results clearly demonstrate the particular advantages of dynamic magnetic resonance in characterizing LCPs.

---

* Present address: BASF Aktiengesellschaft, D-6700 Ludwigshafen, West-Germany

1 Introduction

Thermotropic liquid crystal polymers (LCPs) are of considerable current interest, because of their theoretical and technological aspects [1–3]. Evidently, a new class of polymers has been developed, combining anisotropic physical properties of the liquid crystalline state with characteristic polymer features. This unique combination promises new and interesting material properties with potential applications, for example in the field of high modulus fibers [4], storage technology, or non-linear optics [5].

  Although considerable effort has centered around the synthetic design and macroscopic behaviour of these systems, relatively little attention has been focused on the dynamic and structural features of the molecular units. Up to now, only limited information is available about molecular order and motion in the different polymer phases [6]. On the other hand, exactly these molecular properties are responsible for the macroscopic appearance of the systems. Apparently, the knowledge of the molecular behavior as a function of the chemical structure or sample history offers a means to a directed design of LCPs with well-defined material properties.

In this connection, the development of appropriate methods, capable of elucidating the molecular properties, presents a major challenge in polymer physics. Among the rare techniques available, dynamic magnetic resonance plays an important role [7]. In particular, the employment of electron [8] and nuclear [9] spin probes has turned out to be very useful in acquiring detailed information about order and dynamics on a molecular level. Moreover, by exploiting different relaxation rates a broad dynamic range can be covered, extending from the Hz to the GHz region, necessary for a comprehensive molecular characterization of the systems [10].

In this article we present a study of the molecular properties of LCPs, employing dynamic magnetic resonance techniques. The second chapter briefly reviews the characteristic features of liquid crystal (LC) mesophases. In particular, the various classes of LCPs are discussed with respect to their chemical structures. The following chapter deals with the various magnetic resonance techniques. First, the experimental basis is presented. Then typical results are given to demonstrate the applicability of the methods. A short introduction to the theoretical background concludes this section.

In the subsequent chapters dynamic magnetic resonance experiments of thermotropic side and main chain polymers are presented. Computer simulations provide the orientational distributions and conformations of the polymer chains and the correlation times of the various motions. The results, referring to all polymer phases, are related to the exceptional material properties of LCPs. The discussion clearly demonstrates the power of dynamic magnetic resonance in characterizing such complex chemical systems.

2 Liquid Crystals

2.1 Mesophases

Liquid crystals present the fourth state of matter, which lies between those of the liquid and crystalline phase. Because of this intermediary position, they are often termed mesophases. Liquid crystals are common in living nature. Their technical applications have placed them in the forefront of interest for·scientists engaged in material research [11–12].

Generally, the crystalline phase is characterized by a long-range positional and orientational order of the molecules in the crystal lattice, leading to anisotropic physical properties. In ordinary, non-mesomorphic systems, these long-range orders are lost simultaneously at the melting point, resulting in an isotropic liquid with only short-range ordering between the molecules. LC compounds, however, exhibit a step-wise decay of the crystalline order with increasing temperature, manifested by the appearance of various stable intermediate or mesophases [13, 14].

From the molecular point of view the formation of LC mesophases requires a pronounced anisotropic shape of the molecules. The two basic structures, schematically depicted in Fig. 1, are characterized either by a rod-like (calamitic) or a disc-like (discotic) shape. Thus, a typical low molecular weight liquid crystal (LMLC) is composed of a rigid mesogenic core (mainly phenyl rings) and flexible pendent groups (alkyl chains of variable lengths). In general, the particular chemical structure of the constituent molecules essentially controls the observed long-range order of the various LC phases.

The most important mesophase structures of calamitic LMLCs are schematically depicted in Fig. 2, referring to nematic (N) and smectic (S) phases. The nematic state is characterized by an approximate alignment of the long molecular axes (orientational order) with respect to a local preferential orientation (director z'). For smectic phases a higher degree of long-range order is observed with an additional arrangement of the molecules in smectic layers (one-dimensional positional order). S_A- and S_C-phases display different orientations of the director axes relative to the layer normal, either

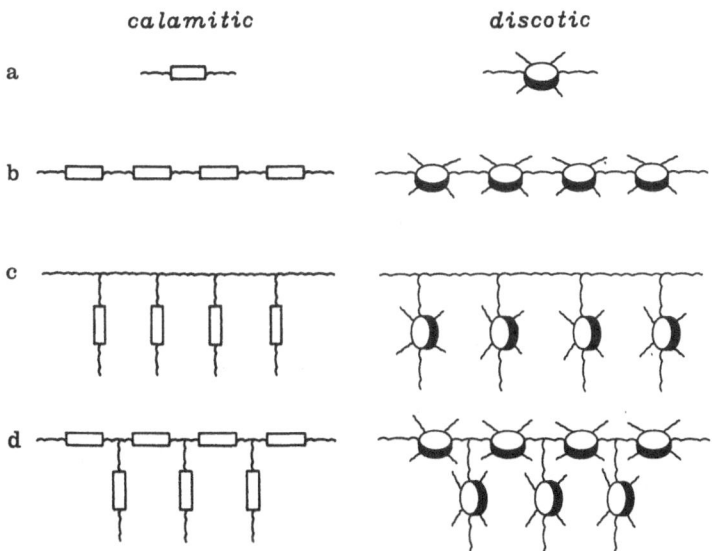

Fig. 1a–d. Molecular structures of rod-like (calamitic) and disc-like (discotic) liquid crystals: **a)** Low molecular weight liquid crystals, **b)** liquid crystal main chain polymers, **c)** liquid crystal side chain polymers, **d)** combined main-chain/side-chain polymers

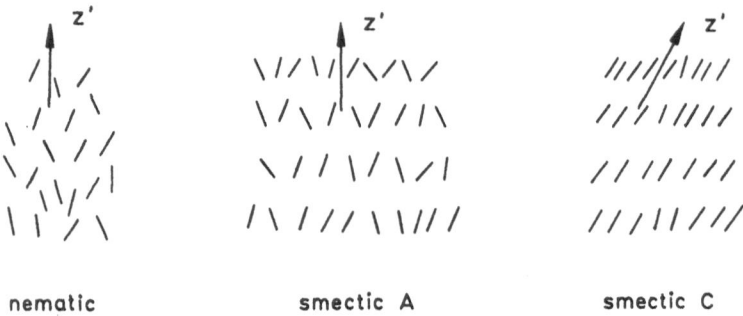

nemotic   smectic A   smectic C

Fig. 2. Schematic representation of various mesophases: Nematic, smectic A and smectic C

parallel (S_A) or inclined at a certain angle (S_C). Various other, higher ordered smectic phases are known (for details see Refs. [14, 15]), but their discussion is outside the scope of this article. Generally, with decreasing temperature, LC mesophases with higher long-range order are formed.

Evidently, these characteristic mesophase structures are responsible for the occurrence of anisotropic physical properties, manifested for example in the magnetic, electrical and optical behaviour [16]. Various successful applications, particularly in the field of electro-optical displays, take advantage of these prominent features, explaining the common interest in LMLCs [12, 14].

2.2 Polymers

Completely new aspects came into this field with the development of LCPs. Generally, conventional non-mesomorphic polymers play an important role in many technological areas, as deduced from their countless applications [17]. In particular, by exploiting the variety in chemical structures polymeric materials with quite different macroscopic properties can be produced. Here, the realization of LCPs offers new and interesting scopes of technological applications, based on the unique combination of specific polymer features with the anisotropic properties of the LC state [18–21].

The synthetic route to LCPs is immediately obvious by recalling the requirements for the appearance of LC mesophases, i.e. the anisotropic shape of the molecules. With this concept in mind two different synthetic routes can be followed, which lead either to LC side chain or main chain polymers. The former systems (see Fig. 1) consist of rigid mesogenic moieties, which are attached to a polymer backbone via flexible spacer [22]. The corresponding chemical constituents, i.e. backbone, spacer and mesogenic groups, have been widely varied. Thus, a large number of LC side chain polymers is now available with different molecular properties, depending on the particular chemical structure [17, 23–27].

On the other hand, LC main chain polymers are prepared by directly incorporating the rigid mesogens in a linear polymer chain [19, 28]. Here, in addition to the chemical constituents, the ratio of flexible to rigid moieties can be varied. Accordingly, a whole range of polymers, extending from totally rigid to completely flexible ones, has been developed, exhibiting a great diversity in the corresponding bulk properties [18–21].

During the last years most efforts at preparing LCPs have focused on the use of rod-like mesogens. Only very recently has the synthesis of mesomorphic polymers with disclike units (see Fig. 1) been reported [29, 30]. Likewise, combined main-chain/side-chain polymers (see Fig. 1), comprising structural aspects of both types of LCPs, have only recently become known [31].

The typical LCP character, resulting from the combination of LC and polymeric features, is primarily reflected in the phase behaviour. In analogy to LMLCs, LCPs exhibit various mesophases (see Fig. 2), often broadened in comparison to those of the monomeric systems. Likewise, the polymeric nature also dominates the behaviour in the solid phase below the anisotropic melt. Here, semicrystalline or glassy states are reported, which are very common for conventional non-mesomorphic polymers.

Moreover, new and exceptional material properties arise from this unique combination of LC and polymeric features, implying various technological applications. For example, the ease with which LC main chain polymers can be oriented by shear fields has been employed in the formation of fibres with unusually high tensile strengths and moduli [1–3]. Similarly, LC side chain polymers in the glassy state have been used as storage materials [32]. Further applications in the display technology and in the field of non-linear optics are currently being developed [5, 33].

2.3 Molecular Properties

Apparently, LCPs exhibit a variety of unusual material properties, subject to increased research activities in this field. One major objective of these studies is the evaluation of a correlation between the macroscopic bulk behaviour and the characteristic molecular properties of these systems. The knowledge of this correlation provides the scientific basis for a directed design of LCPs with well-defined material properties.

The relevant molecular properties of LCPs, comprising order and dynamics, are briefly discussed here. Generally, molecular order of these systems can be specified by contributions to three different levels [10], schematically depicted in Fig. 3. The

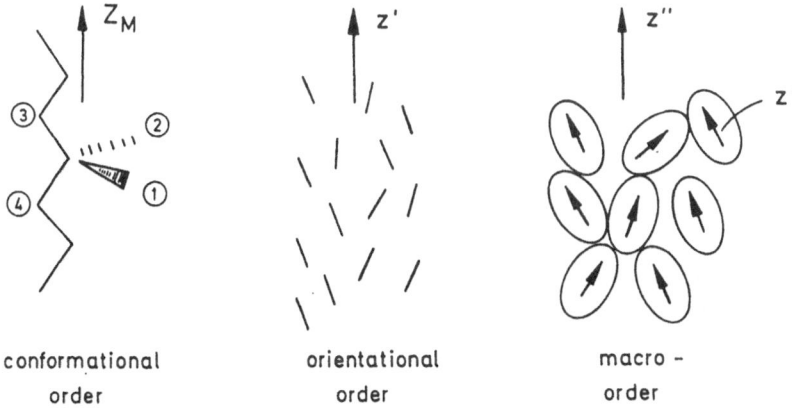

conformational order orientational order macro-order

Fig. 3. Schematic representation of various types of molecular order: Conformational order, orientational order and macroorder

first level accounts for the conformational order of the various alkyl chain segments, characterized by the populations of the four conformational states accessible. The next level considers the orientational order of the polymer chains with respect to a preferential local axis (director z') [34]. Finally, in bulk samples, the macroorder [10, 35] specifies the alignment of the director axes in a laboratory frame (alignment axis z'').

Of course, LCPs are not static but dynamical systems. The most important molecular motions, expected for these systems, are shown in Fig. 4. One can distinguish at least three different motional modes. The intramolecular motions consist of local internal reorientations such as *trans-gauche* isomerization or ring flips [6, 10]. The intermolecular motion is the motion of several chain segments or the molecule as a whole. Two basic motional modes are distinguished, namely rotation about the long molecular axis and reorientation of this axis, respectively [10].

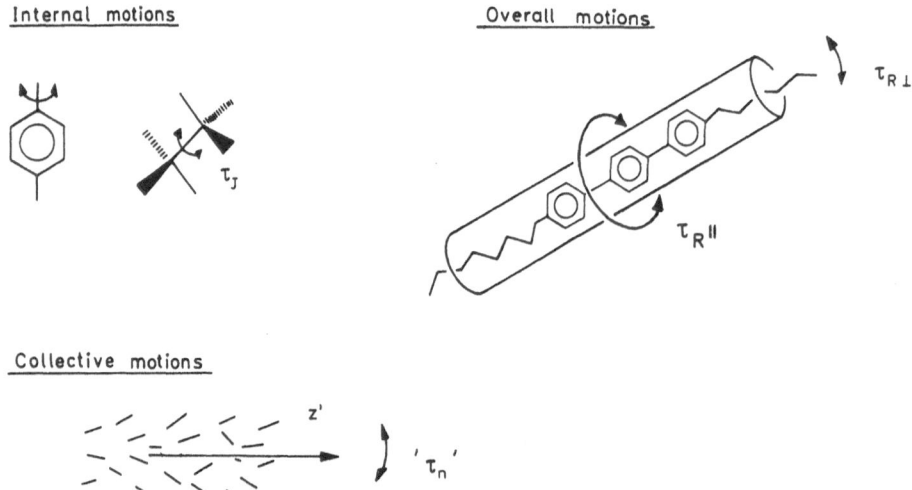

Fig. 4. Schematic representation of various types of motions: Internal motions, overall motions and collective motions

So far, the discussion has been restricted to isolated motions of single molecules. In LCPs, however, collective motions of a large number of molecules may occur. For the latter mechanism, known as order director fluctuations, a broad distribution of correlation times is predicted [36, 37]. In contrast to the isolated modes, discussed above, director order fluctuations are expected to occur only in the mesophase of LCPs, but should be completely absent in the solid and glassy state of these systems.

Generally, the various motions discussed cover a broad dynamic range, extending from the Hz to the GHz region [10, 35]. Thus, only a few experimental techniques are capable of following these motions. Among these, dynamic magnetic resonance plays an important role, since it allows for a simultaneous detection of order and dynamics on the molecular level. An introduction to these techniques is presented in the following chapter.

3 Techniques

Dynamic magnetic resonance techniques, comprising ESR and NMR, are well established tools for the molecular characterization of complex chemical systems, such as liquid crystals or polymers [8, 11, 38–41] . Generally, by exploiting the various relaxation rates of a particular spin system, a broad dynamic range can be covered, compatibly matching that of the molecular motions [7, 10, 42]. However, analysis of these experiments in terms of molecular order and dynamics is often hampered by the well-known difficulties in handling the various couplings of a multispin system. In order to simplify the analysis, different courses have been pursued. One possibility is the application of solid-state NMR techniques, such as magic angle spinning [43], double resonance of multipulse experiments [43, 44], designed to remove undesired magnetic couplings. At best, only a single magnetic interaction remains. However, the theoretical description of these experiments is rather complex, explaining their minor importance for dynamical studies.

A further choice, we focus on, is the employment of electron and nuclear spin probes with *isolated* magnetic interactions. They are introduced into the system of interest, either physically via an external guest molecule (probe molecules) or chemically by a covalent bond at a specific site of the compound investigated (probe nuclei). For the studies, presented in this article, only nitroxide [8] and deuteron [9] spin probes have been employed.

Representative examples are depicted in Fig. 5. The cholestane spin probes (CSL), matching the dimension of most mesogenic units, are expected to reflect the orientational order and dynamics of the LCs. Likewise, deuterons, attached to a phenyl ring or alkyl chain segment, are employed to report on the molecular properties of these groups. It should be noted, however, that the technique is easily extended to other spin probes, appropriate for dynamic NMR or ESR investigations.

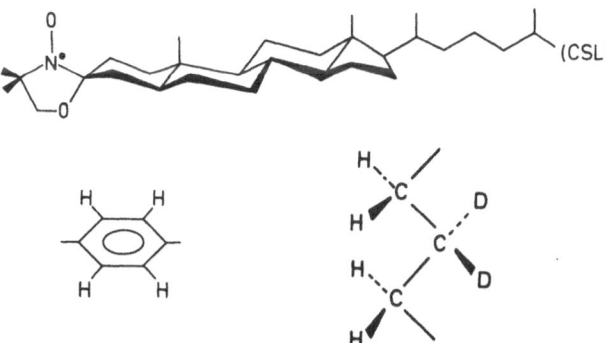

Fig. 5. Molecular structures of the cholestane spin probes (*CSL*) and deuteron spin probes employed

3.1 Dynamic NMR

Generally speaking, dynamic NMR is a time domain technique [45]. The spin system is subject to a sequence of non-selective radio frequency pulses and the response after the last pulse $S(t_2)$ is used to characterize the molecular dynamics of the sample. Fourier transformation (FT) of $S(t_2)$ yields the common frequency spectrum $S(\omega_2)$. Various relaxation times can be evaluated by recording the amplitude of $S(t_2)$ as a function of a particular pulse spacing t_1.

As already mentioned, deuterons (I = 1 spin systems) play an important role as nuclear spinlabels for the molecular characterization of complex chemical and biological systems [9, 38, 46–49]. The specific advantages of these nuclei originate from the dominant quadrupole interactions, axially symmetric along the C–D bond [50]. For I = 1 spin systems there are five independent relaxation times. They can be evaluated by preparing a defined non-equilibrium state of the nuclear magnetization, using a particular pulse sequence and by following the return to equilibrium. Each of these relaxation times defines a specific dynamic window in which molecular motions can be studied. Thus, by combining studies of different relaxation times a broad dynamic range can be covered.

Typical pulse sequences, commonly employed in relaxation studies of I = 1 spin systems are shown in Fig. 6. The quadrupole echo (QE) sequence (top) [51] provides the spin-spin relaxation time T_{2E} [52]. Since T_{2E} is most sensitive to motions with correlation times equal to the inverse quadrupolar coupling constant, quadrupole echo sequences of deuterons (^{2}H) offer a means to study molecular dynamics in the

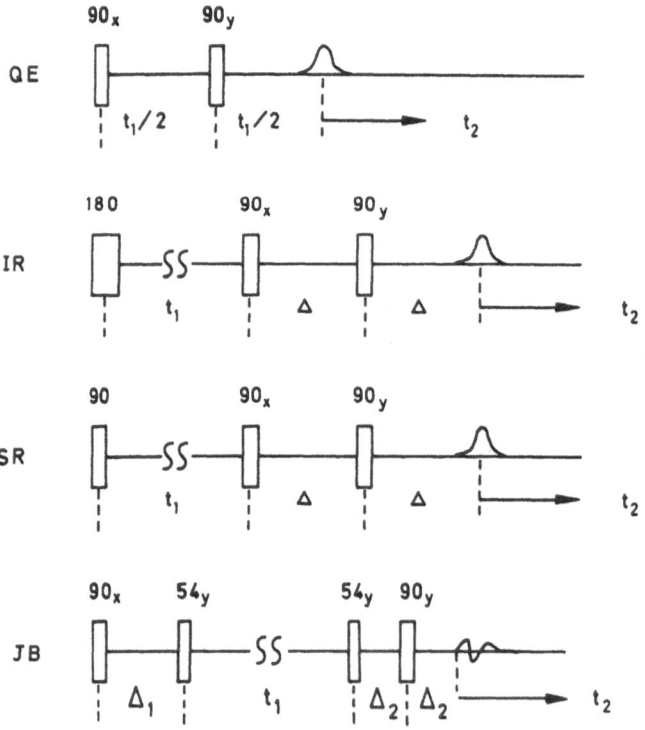

Fig. 6. Schematic representation of various pulse sequences, employed in dynamic NMR of I = 1 spin systems: Quadrupole echo sequence (*QE*), inversion recovery sequence (*IR*), saturation recovery sequence (*SR*) and Jeener-Broekaert sequence (*JB*)

range 10^{-8} s $< \tau_R < 10^{-4}$ s. Faster motions are accessible by employing inversion recovery (IR) or saturation recovery (SR) sequences (center) [45, 53]. From measurements of the signal amplitude as a function of t_1, the spin lattice relaxation times T_{1Z} can be obtained. They are particularly sensitive to motions with correlation times equal to the inverse Larmor frequency. In case of high magnetic fields (B $>$ 5 T) fast molecular dynamics in the range 10^{-12} s $< \tau_R < 10^{-8}$ s can be studied [53]. The range may be extended to slower motions by employing lower relaxation fields (field cycling technique) [54]. Finally, the Jeener-Broekaert (JB) sequence (bottom) [55] permits the study of extremely slow motions. Under certain conditions analysis of the relaxation curve of this sequence yields information about type and timescale of extremely slow motions with correlation times 10^{-4} s $< \tau_R < 10$ s [56]. Thus, by com-

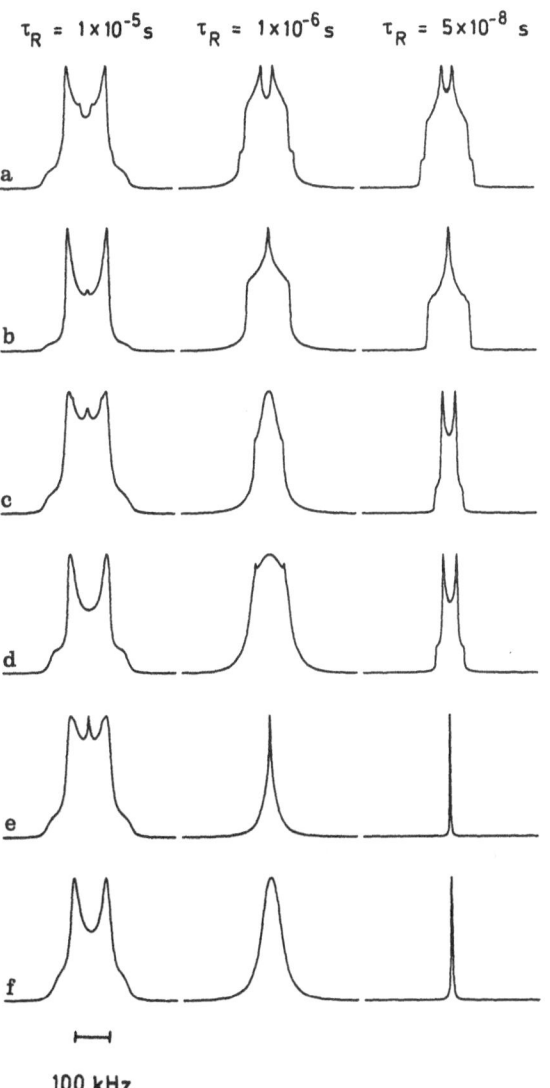

Fig. 7. Dependence of calculated ^{2}H NMR powder spectra on type and timescale of various motions: a) Two-site jumps, $\theta_K = 60°$, b) two-site jumps, $\theta_K = 109°$, c) three-site jumps, $\theta_K = 109°$, d) planar rotational diffusion, $\theta_K = 109°$, e) tetrahedral jumps, f) isotropic spherical diffusion. θ_K = angle between rotation axis and C–D bond direction

bining analyses of quadrupole echo, inversion recovery and Jeener-Broekaert sequences, it is possible to follow dynamic processes over 10–12 orders of magnitude
of correlation times.

The sensitivity of typical ^{2}H NMR experiments to the type and time scale of various
motions is illustrated in Fig. 7. For simplicity, we have chosen equal correlation
times for all motions, namely $\tau_R = 1 \times 10^{-5}$ s, $\tau_R = 1 \times 10^{-6}$ s and $\tau_R = 5 \times 10^{-8}$ s.
The powder spectra refer to quadrupole echo sequences [57], and characterize two-
site jumps (a, b), three-site jumps (c), planar rotational diffusion (d), tetrahedral
jumps (e) and isotropic spherical diffusion (f), respectively. The significant differences
of the lineshapes arise from the different motional anisotropies. Evidently, quadrupole echo spectra [57] contain valuable information on the type of motion [10, 48, 49,
58].

Powder lineshapes obtained via multipulse sequences (see Fig. 6), may be distorted
as compared to those from the free induction decay (FID), due to angular dependent
relaxation times [53, 58, 59]. Generally, one observes that with increasing pulse
separation time t_1 significant spectral changes occur, which are different for the different motions. Thus, the change of the powder lineshape with the pulse separation
is highly indicative of type and timescale of the motion [10, 49, 60–62].

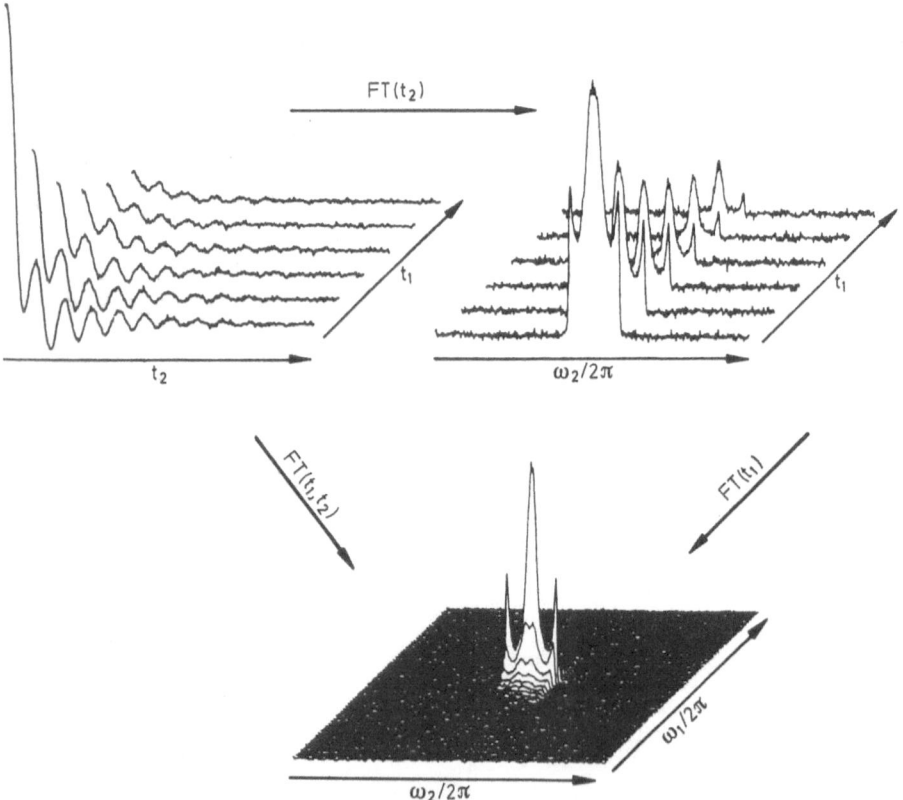

Fig. 8. The formation of two-dimensional NMR relaxation spectra

This is systematically exploited in a two-dimensional (2D) version of the relaxation experiments [63, 64]. The method employs the pronounced anisotropy of the nuclear spin relaxation times, observed for polycrystalline or multidomain samples. Here, the relaxation times vary as a function of the crystal or director orientation, or equivalently as a function of the frequency across the breadth of the powder spectrum. It turns out that this variation is of particular diagnostic importance for the dynamic characterization of molecular systems. Generally, 2D relaxation spectra are obtained by recording the time signals $S(t_2)$ as a function of successive incremented time intervals t_1. A complex FT in both time domains transforms $S(t_1, t_2)$ into a 2D representation $S(\omega_1, \omega_2)$ of the corresponding relaxation experiment, as depicted in Fig. 8 [65].

2D relaxation spectra may thus be regarded as a graph of the relevant natural widths $(1/T_i)$ vs the resonance positions of the dynamical spin packets that constitute the spectrum [66]. For example, cross sections through the 2D quadrupole echo spectrum along ω_1 provide homogeneous linewidths associated with the spin-spin relaxation time T_{2E} [63, 64, 67]. It is found that both the magnitude of T_{2E} and the way in which T_{2E} changes across the spectrum are very dependent upon the character of the molecular motion, responsible for spin relaxation.

This is illustrated in Fig. 9. The 2D spectra refer to quadrupole echo sequences and characterize two possible reorientation mechanisms of a methyl group (three-site jumps vs continuous diffusion). Drastic spectral differences are observed. Apparently, these 2D relaxation spectra sensitively indicate the type of motion. The same is true for the corresponding normalized contour plots (see Fig. 9). We note that similar 2D spectra can be obtained from inversion recovery or Jeener-Broekaert sequences (see Fig. 6) [68]. Thus, by applying this 2D technique to different pulse sequences, the various motions can be differentiated over an extremely wide dynamic range, extending from the fast-rotational to the ultraslow motional regime. Since the different motions (see Fig. 4) modulate different kinds of molecular order (see Fig. 3) these orders can be differentiated, likewise.

3.2 Dynamic ESR

Dynamic ESR spectroscopy, employing free radical spin probes, has been widely used to study the molecular properties of LCs [8, 69, 70]. In particular, nitroxide probes (see Fig. 5) have proven to be ideally suited for this purpose [71-74]. One important reason is their exceptional thermal stability. Moreover, the large anisotropy of the nitrogen hyperfine (HF) interaction [75] makes them sensitive probes for the molecular environment. With some computational effort quantitative details about order and motion can be extracted from the ESR experiments.

The magnitude of the interactions in ESR, exceeding that in NMR by several decades, makes pulsed excitations rather difficult. Consequently, most dynamic ESR studies are performed as continuous wave (CW) experiments, employing a sequential excitation of the spin system. It can be shown, however, that the frequency spectra from the CW technique are identical with those obtained by FT of the FID, following a single pulse [45, 65, 76].

Consequently, CW ESR lineshapes reflect the spin-spin relaxation time T_2 of the paramagnetic species. Since the HF anisotropy of nitroxides is of the order of 100 MHz,

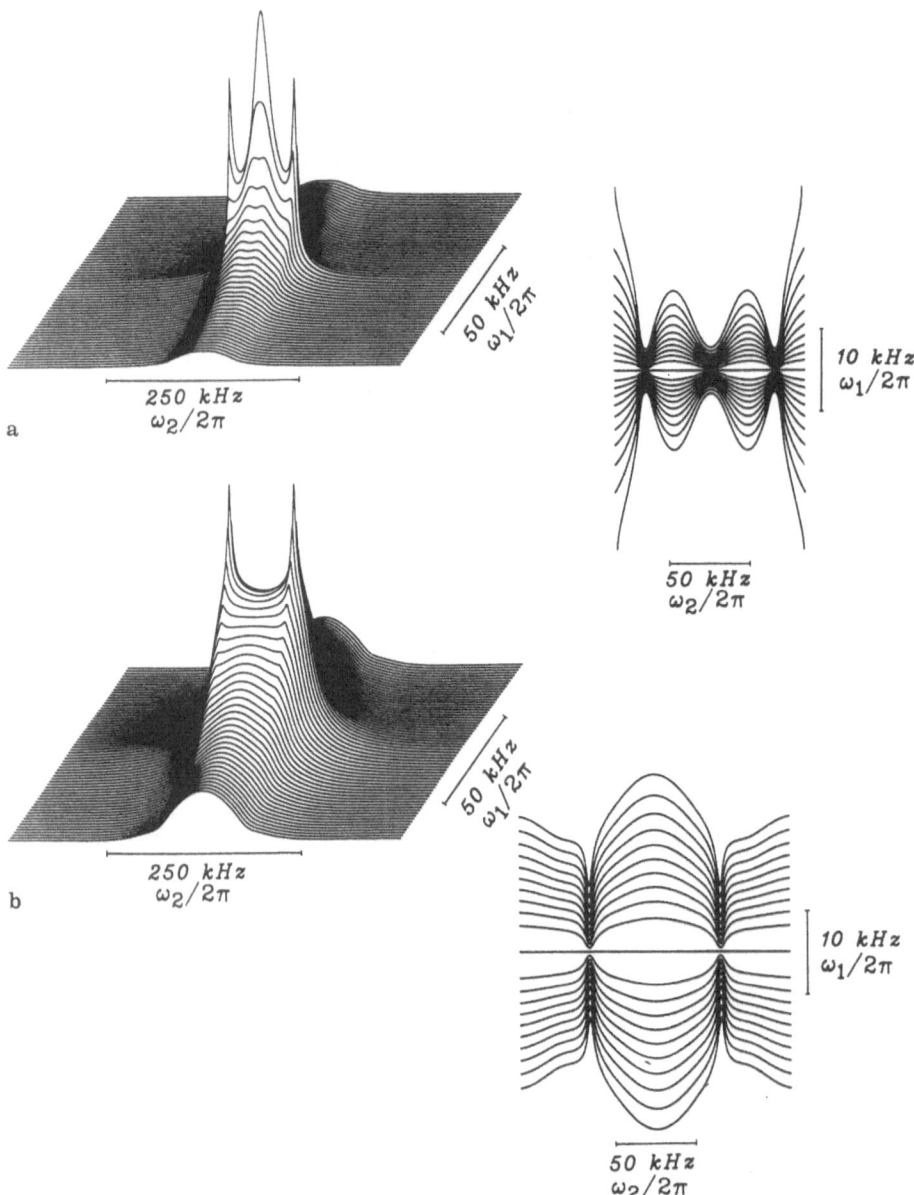

Fig. 9a, b. Calculated two-dimensional NMR relaxation spectra (quadrupole echo sequences) of a solid, containing a rotating (deuteriated) methyl group. **a)** Three site-jumps, $\tau_J = 10^{-6}$ s, **b)** continuous diffusion, $\tau_{R\parallel} = 10^{-6}$ s. Left side: Stack plots. Right side: Normalized contour plots (contours in units of 5% of maximum amplitude)

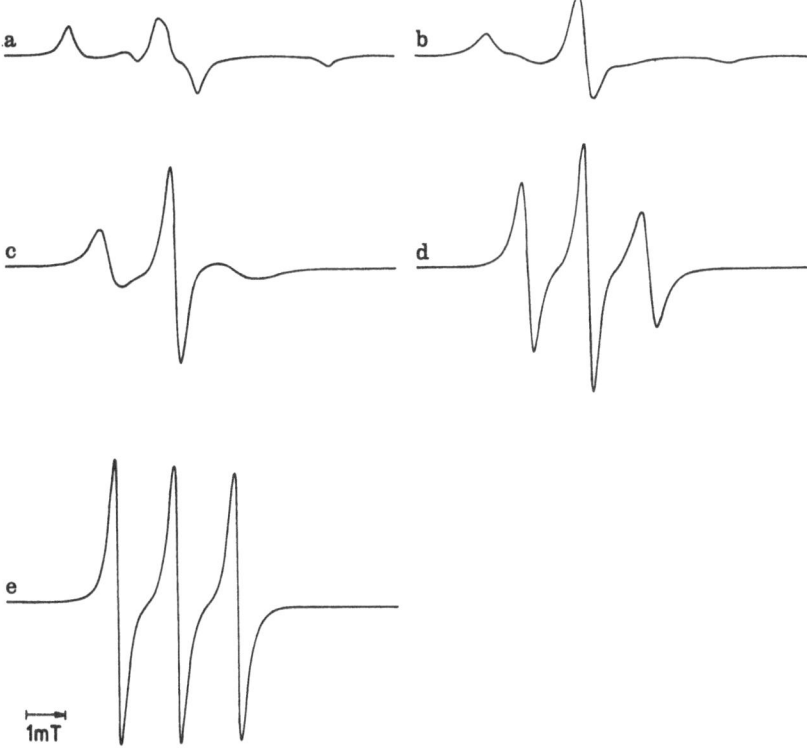

Fig. 10a–c. Calculated ESR spectra of cholestane spin probes, undergoing rotational diffusion in an isotropic medium at various correlation times: **a)** $\tau_{R\parallel} = 5 \times 10^{-6}$ s, **b)** $\tau_{R\parallel} = 5 \times 10^{-8}$ s, **c)** $\tau_{R\parallel} = 5 \times 10^{-9}$ s, **d)** $\tau_{R\parallel} = 5 \times 10^{-10}$ s, **e)** $\tau_{R\parallel} = 5 \times 10^{-11}$ s. The constant simulation parameters are summarized in Table 2. Anisotropy ratio: $\tau_{R\perp}/\tau_{R\parallel} = 5$

their ESR spectra are sensitive to motions between 10^{-10} s $< \tau_R < 10^{-6}$ s [8, 35]. This is demonstrated in Fig. 10. The calculated lineshapes refer to CSL probes (see Fig. 5), undergoing rotational diffusion in an isotropic medium. Drastic spectral changes are observed when the motion increases from the rigid limit (a) to the fast-rotational region (e).

Likewise, CW ESR spectra reflect the degree of molecular ordering, as demonstrated in Fig. 11. The lineshapes refer to CSL probes in a LC medium with the director either parallel (left column) or perpendicular (right column) to the magnetic field. One observes that the angular variation of the spectra significantly increases with the degree of order from $S_{zz} = 0.2$ (top row) through $S_{zz} = 0.6$ (center row) to $S_{zz} = 0.9$ (bottom row). Apparently, nitroxides are sensitive spin probes for LC* media.

In principle, pulsed ESR techniques extend the available dynamic range considerably [77]. However, these experiments are confronted with a variety of technical problems. First, since a common ESR spectrum covers a frequency range of several hundred MHz, the uniform excitation is difficult, requiring short microwave pulses of high intensity. Therefore, usually only small sections out of the whole spectral range are excited [66, 77, 78]. Secondly, the short relaxation times, observed in ESR

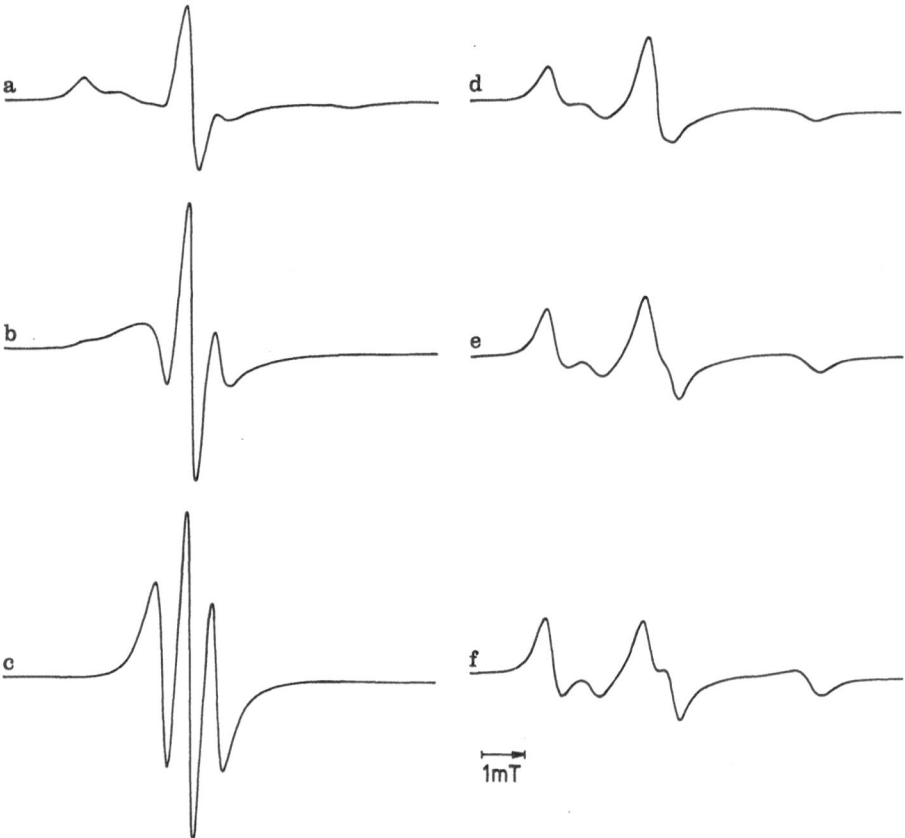

Fig. 11a–f. Calculated ESR spectra of cholestane spin probes, undergoing rotational diffusion in a liquid crystalline medium (monodomain sample) at different order parameters S_{zz}. **a)** $S_{zz} = 0.2$, $\xi = 0°$, **d)** $S_{zz} = 0.2$, $\xi = 90°$, **b)** $S_{zz} = 0.6$, $\xi = 0°$, **e)** $S_{zz} = 0.6$, $\xi = 90°$, **c)** $S_{zz} = 0.9$, $\xi = 0°$, **f)** $S_{zz} = 0.9$, $\xi = 90°$. ξ = angle between director of the liquid crystal and magnetic field. The constant simulation parameters are summarized in Table 2. Motional correlation times: $\tau_{R\perp} = 8 \times 10^{-7}$ s, $\tau_{R\parallel} = 8 \times 10^{-8}$ s

(micro- to nanosecond range), demand for very fast digitizing facilities. Nevertheless, by advances in the microwave and computer technology, various time-domain ESR experiments have recently become feasible.

For example, in electron spin echo (ESE) [78–81] spectroscopy, pulse experiments are performed which are similar to those discussed in the previous section. Again, different relaxation times can be used to extend the dynamical range, accessible by conventional CW techniques. There is, however, one major drawback of dynamic ESE, namely the spectrometer dead time, which prevents the study of fast motions.

The transient ESR technique [82, 83], recently employed in dynamic studies [84], does not suffer from this deficiency. Apparently, the decay of the transient magnetization is determined either by the spin lattice or the spin-spin relaxation time, depending on the motional state of the spin probes [84]. It is this dependence which makes transient ESR such a valuable technique for studying molecular motions over an extremely broad dynamic range. At slow motions the method is limited by the occurrence of dominant T_2 processes other than molecular reorientation. A fast motion limit is reached when T_1 becomes comparable to the time resolution of the detection system. Thus, contrary to ESE, transient ESR is also suitable for studying fast motions, not yet detectable by pulsed ESR because of spectrometer dead time [84].

3.3 Theoretical Background

Analysis of the dynamic magnetic resonance experiments is achieved by using the density operator formalism, outlined elsewhere [10, 35, 49]. Here we summarize important features of this treatment and introduce the simulation parameters. The spin Hamiltonion $H_K(\Omega)$, representing Zeeman, quadrupole or hyperfine interactions (index K) in the laboratory frame x, y, z is conveniently written as [43]

$$H_K(\Omega) = H_K^0 + \sum_{M=-2}^{2} (-1)^M F_K^{(2,-M)}(\Omega) \, T_K^{(2,M)} \tag{1}$$

where H_K^0 is the scalar part and $T_K^{(2,M)}$, $F_K^{(2,-M)}$ denote laboratory frame spin operators and spatial operators, respectively. Generally, nonsecular or pseudosecular terms, $T_K^{(2,M)}$ with $M \neq 0$, have to be retained in the analysis. The orientation dependence of the spatial operators can be evaluated by a threefold transformation from the magnetic axis system X_K, Y_K, Z_K in which the relevant interaction tensor is diagonal ($F_K^{'(2,\pm1)} = 0$). In the first step we transform to a molecular coordinate system X_M, Y_M, Z_M, using the Wigner rotation matrix $D^{(2)}$ ($\phi_K\theta_K\psi_K$). In the second and third step we rotate by the Euler angles ($\phi_M\theta_M\psi_M$) and ($\Phi\Theta\psi$) into the laboratory frame (see Fig. 12) and obtain

$$F_K^{(2,M)}(\Omega) = \sum_{M',M'',M'''=-2}^{2} D_{M'M}^{(2)}(\Phi\Theta\Psi) \, D_{M''M'}^{(2)}(\phi_M\theta_M\psi_M)$$

$$\times \, D_{M'''M''}^{(2)}(\phi_K\theta_K\psi_K) \, F_K^{'(2,M''')} \tag{2}$$

In order to describe the time evolution of the density matrix $\varrho(t)$ during some arbitrary pulse sequence, we divide the sequence into regions, where a pulse is present and regions where there is no pulse. The action of the different non-selective pulses (including a single 90° pulse for the FID which after FT yields the CW frequency spectrum) is considered by unitary transformations employing Wigner rotation matrices [10, 49]. After the pulse the density matrix is assumed to obey the stochastic Liouville equation [85, 86]

$$\frac{\partial}{\partial t} \varrho(\Omega, t) = -(i/\hbar) \, H^X(\Omega) \cdot \varrho(\Omega, t) - \Gamma_\Omega \cdot [\varrho(\Omega, t) - \varrho_{eq}(\Omega)] \tag{3}$$

which we solve using a finite grid point method [87, 88]. Here $H^X(\Omega)$ denotes a superoperator [89] associated with the spin Hamiltonian

$$H(\Omega) = \sum_K H_K(\Omega) \tag{4}$$

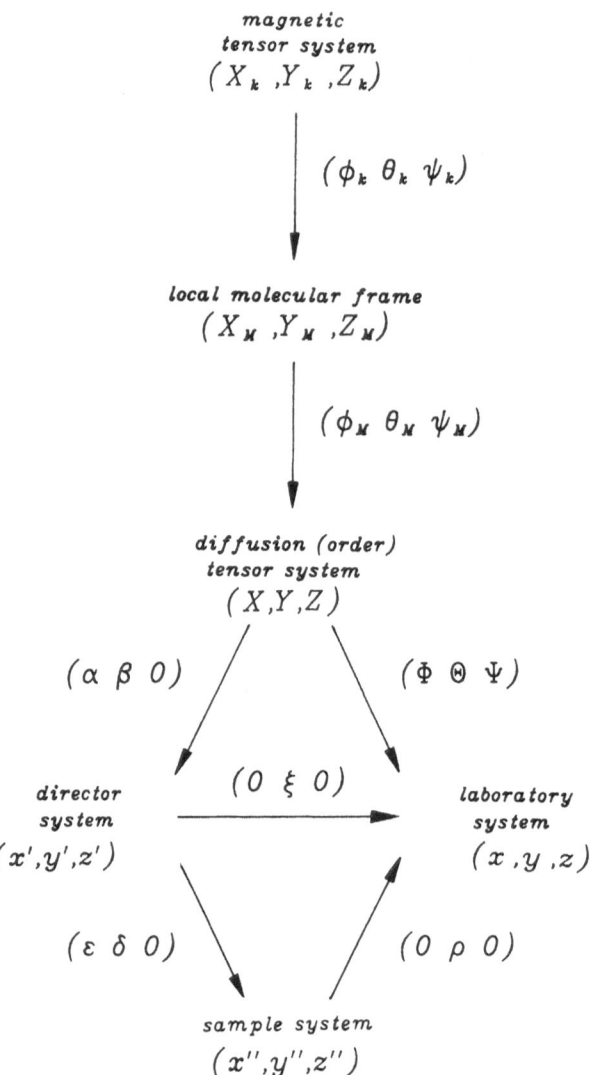

Fig. 12. Notation for coordinate systems and Euler transformations used in the NMR and ESR relaxation model

of the corresponding system (nitroxide or deuteron spin probe). Γ_Ω is the stationary Markov operator for the various rotational processes and ϱ_{eq} is the equilibrium density matrix.

In the finite grid point method [87, 88], the Markov operator is represented by a matrix $W(\Omega, \Omega)$ whose elements give the transition rates between discrete sites of Ω. The values of the transition rates depend upon the model used to describe the motion. For the intramolecular dynamics such as *trans-gauche* isomerization or ring flips (see Fig. 4) a random jump process is assumed. Consequently [90]

$$W(\Omega_m, \Omega_n) = (1/\tau_j) \left[P'_{eq}(\Omega_n) - \delta_{mn} \right] , \tag{5}$$

where τ_J, $P'_{eq}(\Omega_n)$ are the average residence time in one conformation, and the occupation probability of a particular conformation, respectively.

In general, there are only four conformational states for a particular aliphatic chain segment [91]. The corresponding populations $P'_{eq}(\Omega_m)$ (m = 1, ... , 4) may be used to set up a segmental order matrix which on diagonalization yields the segmental order parameters $S_{Z'Z'}$ and $S_{X'X'} - S_{Y'Y'}$ [49]. They express the ordering of the segmental axis Z' and the anisotropy of that order, respectively.

For the intermolecular motion (overall reorientation), rotation through a sequence of infinitesimally small angular steps is assumed. In that case the elements of W(Ω, Ω) must satisfy the following equations [10, 49, 88]

$$W(\Omega_m, \Omega_{m+1}) + W(\Omega_m, \Omega_{m-1}) = (3\Delta^2\tau_R)^{-1} , \qquad (6)$$
$$W(\Omega_m, \Omega_n)\, P''_{eq}(\Omega_m) = W(\Omega_n, \Omega_m)\, P''_{eq}(\Omega_n) ,$$
$$W(\Omega_m, \Omega_m) = -(3\Delta^2\tau_R)^{-1} ,$$

where Δ is the angular separation of adjacent grid points. Solving these equations, one can establish values for all intermolecular transition rates in terms of two rotational correlation times $\tau_{R\perp}$ and $\tau_{R\|}$ and the equilibrium population $P''_{eq}(\Omega)$ of the orientational sites; $\tau_{R\perp}$ is the correlation time for reorientation of the symmetry axis of the diffusion tensor, while $\tau_{R\|}$ refers to rotation about it (see Fig. 4).

The equilibrium population, $P''_{eq}(\Omega_m)$, of a particular site is related to the orientational distribution function [92]

$$f(\Phi\Theta\Psi) = N_1 \exp\{A_{00}D^{(2)}_{00}(0\beta0) + A_{20}[D^{(2)}_{20}(\alpha\beta0) + D^{(2)}_{-20}(\alpha\beta0)]\} \qquad (7)$$

by an integration over the area of that site. In expressing the Wigner elements $D^{(2)}_{M0}(\alpha\beta0)$ as function of $\Phi\Theta\Psi$ we assume that the order tensor is colinear with the diffusion tensor (see Fig. 12). The coefficients A_{00} and A_{20} characterize the orientation of the molecules with respect to a local director z' while the angle ξ (see Fig. 12) specifies the orientation of z' in the laboratory frame. The orientational order parameters S_{XX}, S_{YY} and S_{ZZ} are related to the coefficients A_{00} and A_{20} by mean-value integrals [92].

In unoriented systems, the director axes are randomly distributed. In macroscopically ordered samples, however, z' can be specified with respect to a sample system x'', y'', z'', generally defined by the alignment experiment used to prepare the sample (see Fig. 12). Of course, all director axes need not have the same orientation; instead they may be distributed according to the probability function [35]

$$f(\epsilon\delta0) = N_2 \exp\{B_{00}D^{(2)}_{00}(0\delta0) + B_{20}[D^{(2)}_{20}(\epsilon\delta0) + D^{(2)}_{-20}(\epsilon\delta0)]\} \qquad (8)$$

where the parameters B_{00} and B_{20} specify the orientation of the director axes in the sample system (see Fig. 12). Note that function (8) allows for a biaxial director distribution, which may occur for example in simultaneously applied electric and magnetic fields [35]. For convenience the coefficients B_{00} and B_{20} are transformed into macroorder parameters $S_{x'x'}$, $S_{y'y'}$ and $S_{z'z'}$ by evaluating the corresponding mean-value integrals [35, 92].

3.4 Experimental Section

Materials. The LCPs **1–4** and low molecular weight analogues **5–8** which we consider in detail have the molecular structures shown in Fig. 13. The Greek letters, α, β, δ, φ_C and φ_0, refer to five different derivatives of the same LC, deuteriated at different sites of the mesogen or alkyl chain, as indicated in the formula of LCP **4**. Synthetic

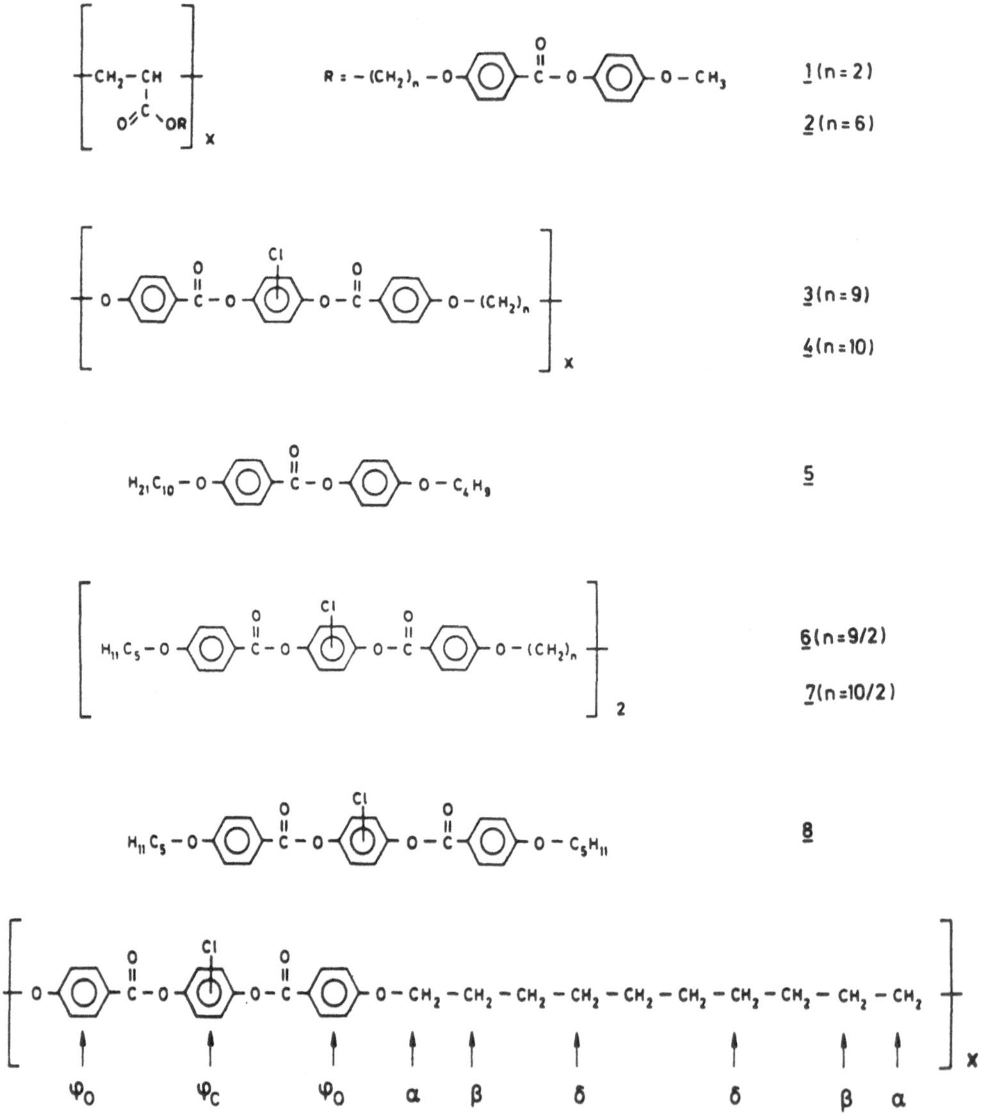

Fig. 13. Molecular structures of the liquid crystal systems studied: Liquid crystal side chain polymers (**1, 2**), liquid crystal main chain polymers (**3, 4**) and low molecular weight analogues (**5–8**). The Greek letters, α, β, δ, φ_C and φ_0, refer to five different derivatives of the same liquid crystal, deuteriated at different sites of the mesogen or alkyl chain, as indicated in the formula of liquid crystal polymer 4

Table 1. Physical properties of the liquid crystal systems studied

Liq. cryst. system	Spacer length	Mol. wt.[a] $\overline{M}_n$	Phase trans. temperatures[b] T_{kl}/K	Clearing enthalpy[b] $\Delta H_{ni}/(kJmol^{-1})$	Clearing entropy[b] $\Delta S_{ni}/(Jmol^{-1}K^{-1})$
LCP 1	2	4500	g 328 n 374 i	0.5	1.3
LCP 1	2	14000	g 335 n 388 i	0.6	1.5
LCP 1	2	21000	g 334 n 388 i	0.7	1.8
LCP 2	6	14000	g 302 s_A 367 n 393 i	0.9	2.3
LCP 3	9	12000	g 303 c 429 n 543 i	6.5	11.8
LCP 4	10	12000	g 303 c 430 n 552 i	6.7	12.1
LCP 4	10	30000	g 313 c 450 n 567 i	7.1	12.4
LMLC 5	—	426	c 326 s_C 329 s_A 352 n 361 i	1.6	4.5
LMLC 6	9	1034	c 377 n 508 i	3.1	6.1
LMLC 7	10	1048	c 403 n 508 i	4.1	8.0
LMLC 8	—	525	c 369 n 454 i	2.1	4.6

[a] Determined by vapour pressure osmometry.
[b] Determined by polarizing microscopy and differential scanning calorimetry; g = glassy, c = crystalline, s_C = smectic C, s_A = smectic A, n = nematic, i = isotropic.

procedures are described elsewhere [28, 93–95]. The relevant physical properties of the LCs, evaluated by standard methods (vapour pressure osmometry, differential scanning calorimetry, polarizing microscopy) are listed in Table 1. For further details we refer to Refs. [93–95]. Note, however, that the side chain LCPs **1** and **2** exhibit homogeneous solid phases (glass transition), in contrast to the main chain LCPs **3** and **4**, for which a semicrystalline behaviour is observed. The cholestane spin probe CSL (see Fig. 5), employed in the ESR investigations, was purchased from Syva and used without further purification.

ESR Measurements. The side chain LCPs **1** and **2** and the corresponding low molecular weight analogue **5** were studied by dynamic ESR techniques [35]. The weight fraction of the CSL spin probe employed was $< 10^{-3}$. Sample cells were constructed of two quartz plates coated with tin dioxide to make them conducting. The thickness of the cell was 250 μm.

Macroscopic alignment of the samples was achieved in the nematic state shortly below the clearing temperature T_{ni} with the quartz plates parallel to a magnetic field of 0.33 and 0.7 T. In addition a high-frequency electric field (50 kHz) of E = 50 kV/cm was applied. The CW ESR measurements were performed on a Varian E9 X-band spectrometer ($\omega_0/2\pi = 9.4$ GHz) using 100 kHz field modulation. The temperature of the sample was controlled by a home-built variable temperature control unit.

NMR Measurements. Specifically deuteriated LC main chain polymers **3** and **4** and corresponding low molecular weight analogues **6–8** have been studied by ^{2}H NMR spectroscopy. Macroscopically aligned samples were obtained in the nematic melt by applying strong magnetic and electric fields [10]. Likewise, solid state extrusion [10] and melt spinning techniques [96] produced highly oriented fibres. Quenched polydomain samples were prepared by rapidly cooling from the anisotropic melt in the absence of any external field. Heat treatment for about 40 min at 15 K below the melt transition yielded the annealed samples discussed in chapter 5.

The ^{2}H NMR experiments were performed on Bruker CXP 300 and MSL 300 pulse spectrometers at $\omega_0/2\pi = 46.1$ MHz (B = 7T) applying the various pulse sequences discussed above (see Fig. 6). Typically, the width for a 90° pulse was 2.0 µs (5 mm coil), employing a home-built probe, equipped with a goniometer. All experiments were recorded using quadrature detection and appropriate phase cycling schemes. The number of scans varied between 500 and 5000 depending on the particular sample and experiment. In the 2D relaxation experiments a typical data set consisted of 64 sampling points in the t_1 domain and of 2048 sampling points in the t_2 domain. The temperature of the samples was controlled by a Bruker control unit, operating in the range 120 K < T < 575 K.

Computations. A Fortran program package, based on the theoretical approach outlined in the previous section, was employed to analyze the ESR and NMR experiments. The programs simulate dynamic magnetic resonance experiments of nitroxide radicals and ^{2}H spin probes undergoing inter- and intramolecular motion in an aniso-

Table 2. Constant Parameters Used in the Calculations of ESR spectra of Cholestane Spin Probes in Liquid Crystal Side Chain Polymers and Low Molecular Weight Analogues

Liq. cryst. system	g-Tensor[a]			Hyperfine tensor[a]			Magnetic tensor orient.[b]		
	$g_{X_KX_K}$	$g_{Y_KY_K}$	$g_{Z_KZ_K}$	$A_{X_KX_K}$/mT	$A_{Y_KY_K}$/mT	$A_{Z_KZ_K}$/mT	ϕ_K/deg	θ_K/deg	ψ_K/deg
LCP 1	2.0088	2.0063	2.0022	0.60	0.52	3.35	270	270	0
LCP 2	2.0088	2.0063	2.0022	0.58	0.50	3.31	270	270	0
LMLC 5	2.0088	2.0063	2.0022	0.59	0.51	3.29	270	270	0

[a] Diagonal in X_K, Y_K, Z_K.
[b] Euler angles relating magnetic and diffusion (order) tensor systems, since $\phi_M = \theta_M = \psi_M = 0$ in this case (see Fig. 12).

Table 3. Constant parameters used in the analysis of ^{2}H NMR experiments of liquid crystal main chain polymers and corresponding model compounds

Deuteron spinlabel site[a]	Quadrupole coupling constant[b] $(e^2qQ/h)/(kHz)$	Magnetic tensor orientations[c]		Molecular frame orientation[d]	
		ϕ_K/deg	ψ_K/deg	ϕ_M/deg	θ_M/deg
φ_0 (a)	185	60, 60	0,180	0	−12
φ_0 (b)	185	120.5, 120.5	0,180	0	−12
φ_C (a)	185	−61, −61	0,180		−1
φ_C (b)	185	−119.2, −119.2	0,180		−1
φ_C (c)	185	56, 56	0,180		−1
α	165	90, 90, −144.75, −35.25	144.75, 35.25, 90, 90	90	12
β	165	90, 90, 144.75, 35.25	−144.75, −35,25, 90, 90	90	12
δ	165	90, 90, 144.75, 35.25	−144.75, −35.25, 90, 90	90	12

[a] See Fig. 13.
[b] The principal values of the quadrupole coupling tensor Q are given by $Q_{X_KX_K} = Q_{Y_KY_K} = -2Q_{Z_KZ_K}$ $\frac{1}{4}e^2qQ$, since the asymmetry parameter $\eta = 0$.
[c] Euler angles relating magnetic tensor system and local molecular frame (see Fig. 12).
[d] Euler angles relating local molecular frame and diffusion (order) tensor system (see Fig. 12).

tropic medium. The numerical integrations of Eq. (3) are readily achieved using the Rutishauser [97] or Lanczos algorithm [98]. Tables 2 and 3 summarize the constant parameters used in the calculations. They were obtained from an analysis of the corresponding slow-motional (rigid limit) spectra.

4 Liquid Crystal Side Chain Polymers

Molecular order and dynamics of the LC side chain polymers **1, 2** and the low molecular weight analogue **5** (see Fig. 13) have been investigated with dynamic ESR, using the CSL spin probe (see Fig. 5) [35, 99, 100]. In the following representative results are discussed in relation to other studies of these systems. Finally, the observed molecular behaviour is contrasted with the exceptional material properties of LC side chain polymers.

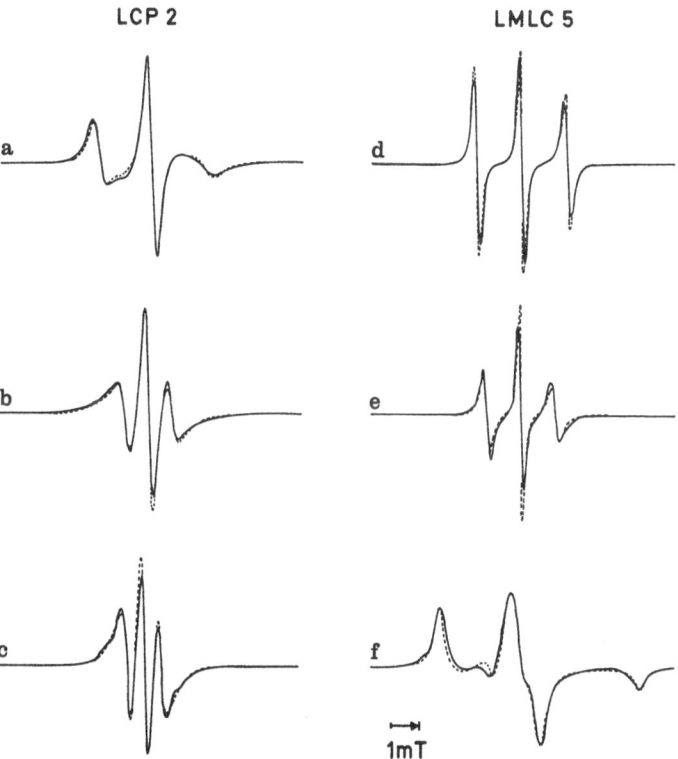

Fig. 14a–f. Experimental (———) and simulated (– – –) ESR spectra of the cholestane spin probes in macroscopically aligned samples ($S_{x'x'} = 1.0$, $S_{z'z'} = -0.5$) of the side chain polymer **2** (*a–c*) and the monomeric analogue **5** (d–f) at three different temperatures in the isotropic (a, d), nematic (b, e), glassy (c) and crystalline (f) phase. The spectra refer to $\varrho = 90°$ and **a)** T = 393 K, **b)** T = 369 K, **c)** T = 263 K, **d)** T = 370 K, **e)** T = 360 K, **f)** T = 263 K. Simulations were obtained with the parameters given in Table 2 and in Figs. 15 and 17. $S_{x'x'}$, $S_{z'z'}$ = macroorder parameters. ϱ = angle between quartz plate normal and magnetic field

4.1 Molecular Dynamics

Previous ESR studies of LMLCs [8, 35, 101, 102] have indicated that the CSL probes are located in the mesogenic core region, reflecting the molecular order and dynamics of these groups. Figure 14 compares ESR spectra of the doped polymer **2** with those of the low molecular weight analogue **5**. The spectra refer to macroscopically aligned samples ($\varrho = 90°$) and three different temperatures, characterizing the isotropic, nematic and solid state, respectively. The upper lineshapes (Fig. 14a, d), characteristic of an isotropic fluid medium, reflect distinct molecular motions in the two systems. Cooling below the clearing temperature causes significant spectral changes. The central spectra (Fig. 14b, e) now indicate oriented LC phases. As one sees, molecular order is essentially retained when the polymer is cooled into the solid state (Fig. 14c) in contrast with the low molecular weight liquid crystal, exhibiting a random solid phase (Fig. 14f) [35].

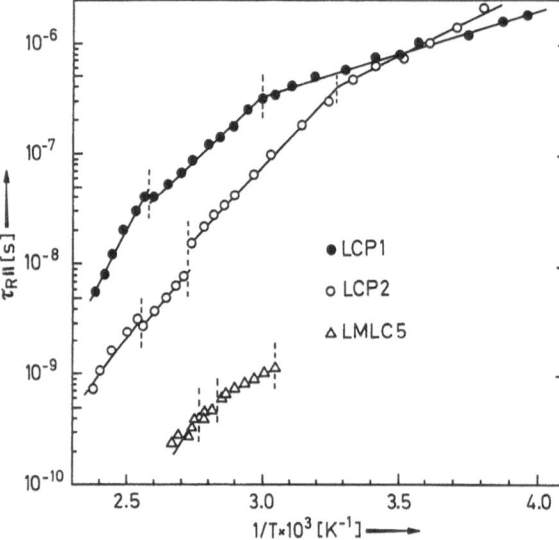

Fig. 15. Arrhenius representation of the rotational correlation times $\tau_{R\parallel}$, characterizing the molecular dynamics of cholestane spin probes in various liquid crystal systems. The correlation times refer to side chain polymer **1** ($\bullet$, $\overline{M}_n = 14000$) side chain polymer **2** ($\bigcirc$, $\overline{M}_n = 14000$) and the low molecular weight analogue **5** ($\triangle$). *Dashed lines* indicate different phase transitions (see Table 1)

Table 4. Dynamic parameters of cholestane spin probes in liquid crystal side chain polymers and low molecular weight analogues: Rotational activation energies and anisotropy ratios

Phase[a]	LCP 1[b]		LCP 2		LMLC 5	
	$E_{R\parallel}/(kJ/mol)$	$\tau_{R\perp}/\tau_{R\parallel}$	$E_{R\parallel}/(kJ/mol)$	$\tau_R/\tau_{R\parallel}$	$E_{R\parallel}/(kJ/mol)$	$\tau_{R\perp}/\tau_{R\parallel}$
i	89.4	7	67.3	7	53.8	5
n	45.4	10	58.1	10	46.7	10
s_A	—	—	50.1	10	25.9	25
g	15.2	10	24.9	10	—	—

[a] i = isotropic, n = nematic, s_A = smectic A, g = glassy.
[b] $\overline{M}_n = 14000$

Computer simulations of the experimental spectra, applying the theoretical approach discussed above, provided detailed information about the spin probe motions in the various systems [35]. In Fig. 15 the correlation times $\tau_{R\parallel}$ for long axis rotation are plotted as a function of $1/T$. They refer to LCP **1** ($\overline{M}_n = 14\,000$, full circles), LCP **2** ($\overline{M}_n = 14\,000$, open circles) and the model compound **5** (open triangles). It should be noted that $\tau_{R\parallel}$ varies by four orders of magnitude (0.2 ns $\leq \tau_{R\parallel} \leq$ 2000 ns), reflecting the complex molecular dynamics of side chain polymers and low molecular weight analogues in the isotropic, liquid crystal, and glassy state.

Inspection of the logarithmic plots reveals several discontinuities, which occur at the phase transitions. Within a particular phase the plots are linear. From the slopes of the straight lines the rotational activation energies $E_{R\parallel}$ have been determined. They are listed in Table 4. As one can see, $E_{R\parallel}$ decreases with increasing order of the phase. Note the surprising low activation energies 15 kJ/mol $\leq E_{R\parallel} \leq$ 25 kJ/mol for the glassy state of the polymers. The anisotropy ratios $\tau_{R\perp}/\tau_{R\parallel}$, also listed in Table 4, remain constant within a phase, varying from $\tau_{R\perp}/\tau_{R\parallel} = 5$ to $\tau_{R\perp}/\tau_{R\parallel} = 25$. Generally, these ratios are consistent with the geometry of the probe ($\tau_{R\perp}/\tau_{R\parallel} = 4.7$) only in the isotropic phase [73, 101].

The most striking feature of Fig. 15 is the large difference in correlation times of polymers and low molecular weight analogues. Closer inspection reveals that $\tau_{R\parallel}$ in LCP **1** (full circles) is 300 times larger than $\tau_{R\parallel}$ in LMLC **5** (open triangles), when referred to the same temperature in the nematic phase [8, 35, 74, 100]. This enormous slowdown of $\tau_{R\parallel}$ indicates a strong coupling of main and side chain motions via the short spacer. As expected, this dynamic coupling is weaker in the case of polymer **2** (open circles), having the long spacer [22, 23]. Note, however, that the correlation times in LCP **2** and LMLC **5** still differ by a factor of 30. Thus decoupling of the mesogenic groups from the main chain, while effective, is not complete.

In this connection, the effect of the molecular weight $\overline{M}_n$ on the side chain motion has been studied [35]. Variation of $\overline{M}_n$ between 4500 $\leq \overline{M}_n \leq$ 21 000 produced only minor changes of the correlation times, in contrast to detectable changes of the thermodynamic parameters (see Table 1) [103–105]. Apparently, the plateau values for the side chain motions are already reached at a low degree of polymerization ($\overline{M}_n \leq$ 4500).

An interesting feature of Fig. 15 is the break of the curves at the calorimetric glass transition T_g [39, 100, 106, 107], indicating a change in the microscopic molecular dynamics. However, because of the high frequency ESR measurements the break cannot be assigned to the glass transition process [6]. Rather, the CSL probes reflect the occurrence of a secondary motion involved in the side group dynamics. In the absence of additional experiments, however, an unequivocal assignment is not yet possible.

Complimentary ^{2}H NMR studies of the glassy state of **1** and **2** have recently appeared [108–110]. Apparently, 180° jumps of the phenyl rings present the dominant process in this phase. A distribution of correlation times has been determined for this motion, extending over 2.2 and 2.6 decades [110]. The activation energies range from 42 kJ/mol to 46 kJ/mol in agreement with results obtained for conventional [7, 40, 111–113] and combined main-chain/side-chain LC polymers [114]. In the latter study the dynamics above T_g, including overall rotation, has also been evaluated [115].

A variety of LC side chain polymers has been characterized by dielectric studies

[116–122]. In general, several different relaxation processes are observed and assigned to specific motional modes. Typically, the α-process is associated with a reorientation of the polymer backbone (glass transition process), while the β-, β'-, γ- and δ-processes refer to overall and local motions of the side groups. Interestingly, the β'-relaxation occurs in the same temperature and frequency range as observed for ring flips in ^{2}H NMR [110]. Generally, however, the discussed γ- and δ-processes are not yet compatible with NMR or ESR results. Additional comparative studies are required to clarify this point.

Generally, there is a complex relation between the rotational correlation times of a spin probe and the viscosity of the surrounding matrix, depending upon the nature of the rotor-matrix interactions [35]. Presuming that these interactions are the same in the side chain polymers 1, 2 and the low molecular weight analogues 5 we expect the rotational viscosities γ_1 of the LC polymers to be at least a factor of 30 (long spacer) or 300 (short spacer) larger than those of the corresponding monomers under the same conditions (see Fig. 15). These predictions are in qualitative agreement with recent measurements of γ_1 for several side chain polymers [105, 123–126].

The observed γ_1 values, exceeding those of LMLCs by orders of magnitude, are of considerable interest for the use of these polymers in display and storage devices. It has been shown that the electro-optical response times are proportional to γ_1, implying that LC side chain polymers are not suited for fast-switching displays [127]. There are, however, other promising applications of these systems in the storage technology, as we shall see in the following sections.

4.2 Molecular Order

Orientational order of LCPs is conveniently described in terms of two different order parameters, characterizing the average orientation of the mesogenic units within a molecular domain (microorder) and the macroscopic alignment of the domains (see Fig. 3) [10, 35]. The measurement of these parameters is important for relating macroscopic physical properties of LCPs to their molecular structure. Various methods have been used to measure order parameters in these systems, including the use of X-ray diffraction [128–130], birefringence [103, 104], and linear dichroism [104, 131], but these methods can not essentially separate the two types of orientational order.

Dynamic magnetic resonance techniques do not suffer from this deficiency. In particular, the use of CSL probes [8, 101, 102] offers the chance of determining the micro- and macroorder unequivocally. This is demonstrated in Fig. 16. The spectra refer to the side chain polymers 2 ($\overline{M}_n = 14000$) and two different temperatures (T = 382 K, left column; T = 263 K, right column). Drastic spectral changes are observed when the sample is rotated (top row, $\varrho = 0°$; central row, $\varrho = 50°$, bottom row, $\varrho = 90°$). Comparison of the angular variation in the nematic (left column spectra) and supercooled smectic phase (right column spectra) reveals the crucial effect of the microorder on the spectral feature.

In Fig. 17 the microorder parameters S_{ZZ} [34] of the CSL probes in the various systems are plotted as a function of the reduced temperature $T^* = T/T_{ni}$ [35, 99, 100]. They refer to LCP 1 ($\overline{M}_n = 14000$, full circles), LCP 2 ($\overline{M}_n = 14000$, open circles) and the low molecular weight analogue 5 (open triangles). One sees that in the iso-

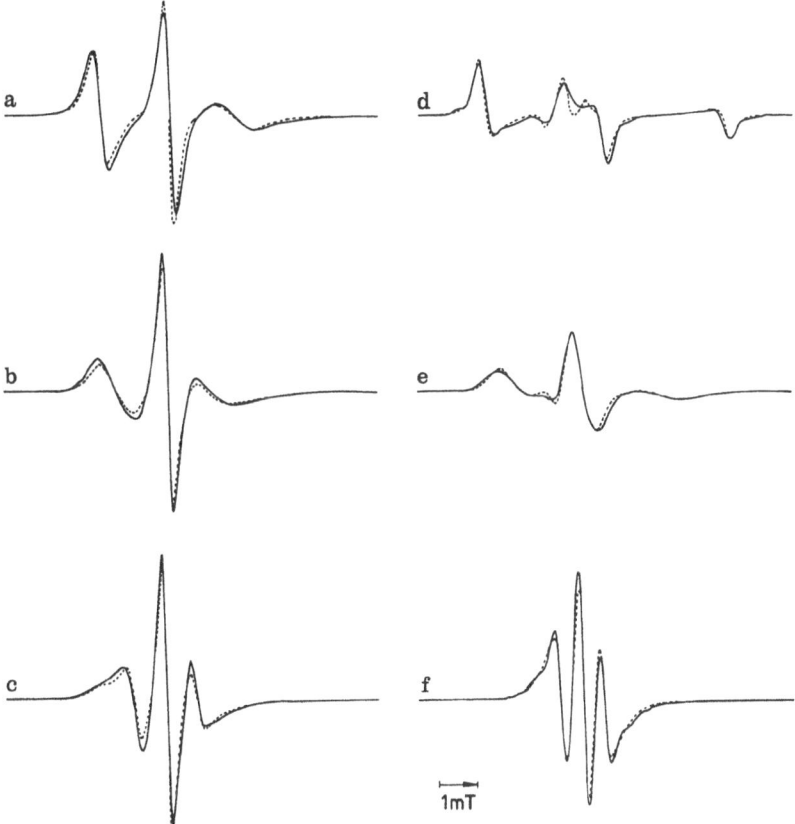

Fig. 16a–f. Experimental (————) and simulated (————) ESR spectra of the cholestane spin probes in macroscopically aligned samples ($S_{x'x'} = 1.0$, $S_{z'z} = -0.5$) of the side chain polymer **2** at T = 382 K (left column, nematic phase) and T = 263 K (right column, glassy smectic state). The spectra refer to three different orientations ϱ of quartz plate normal and magnetic field, namely $\varrho = 0°$ (**a, d**), $\varrho = 50°$ (**b, e**) and $\varrho = 90°$ (**c, f**). Simulations were obtained with the parameters given in Table 2 and in Figs. 15 and 17. $S_{x'x'}$, $S_{z'z'}$ = macroorder parameters

tropic phase $S_{ZZ} = 0$, indicating a random orientation of the spin probes. At the isotropic-nematic transition ($T^* = 1.0$) all order parameters jump to finite values, which depend on the system studied. Generally, side chain polymers exhibit lower order parameters $((S_{ZZ})_{ni} = 0.30)$ than the monomeric analogues $((S_{ZZ})_{ni} = 0.45)$, in agreement with the corresponding clearing entropies ($\Delta S_{ni} \sim 2\,\mathrm{Jmol^{-1}\,K^{-1}}$ vs $\Delta S_{ni} = 4.5\,\mathrm{Jmol^{-1}\,K^{-1}}$), listed in Table 1 [11, 93].

Lowering the temperature increases S_{ZZ} for all systems to the same limiting value of $S_{ZZ} \sim 0.65$ in the nematic phase [35, 99, 100]. Note, that the order parameter curve of polymer **1** exhibits a horizontal slope at the glass transition. A further jump of S_{ZZ}, observed for the polymers **2**, indicates an additional smectic A phase with a limiting value of $S_{ZZ} \sim 0.9$. No such discontinuity is detected for the monomeric liquid crystal **5**, exhibiting a smectic A phase, likewise. Comparison of the order parameters of Fig. 17 with those of ^{2}H NMR [108–110] and birefringence studies

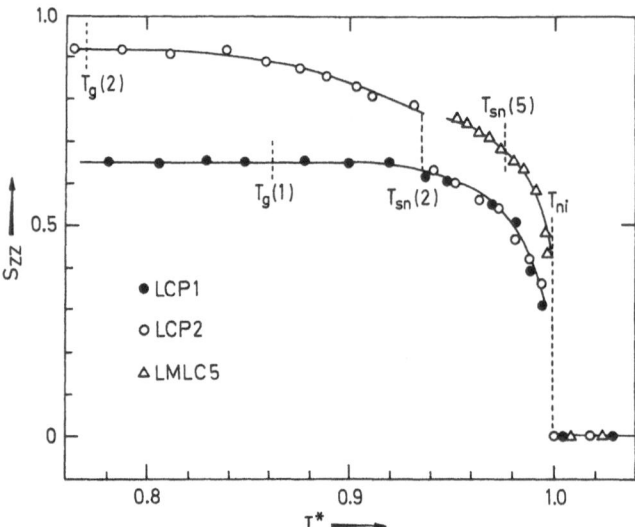

Fig. 17. Temperature dependence of the orientational order parameters S_{zz} of the cholestane spin probes in various liquid crystal systems. The order parameters refer to side chain polymer **1** ($\bullet$, $\overline{M}_n = 14000$), side chain polymer **2** ($\bigcirc$, $\overline{M}_n = 14000$) and the low molecular weight analogue **5** ($\triangle$). *Dashed lines* indicate different phase transitions (see Table 1). $T^* =$ reduced temperature

[103, 104] shows that the CSL probes reliably reflect the microorder of the mesogenic groups.

Surprisingly, the limiting value of the nematic order parameter of the polymers ($S_{zz} \sim 0.65$), independent of spacer length ($n = 2,6$) and molecular weight ($4500 < \overline{M}_n < 21000$) [35], is as high as that of the monomeric analogue. Linkage of the mesogenic units to the polymer backbone does apparently not restrict their orientational order, even in the case of a short spacer. It is evident, however, that in the low molecular weight system the order parameter exhibits a much stronger temperature dependence than in the polymeric one. This, of course, reflects the wider temperature range of the polymers, incidently increasing with molecular weight. Above a certain degree of polymerization, however, the nematic range remains constant (see Table 1), according to a plateau effect, observed for other side chain polymers, likewise [103, 104].

Apparently, there is a decoupling of side and main chain order via flexible spacers, as predicted by the spacer model [22]. According to this model the polymer backbone should retain its random coil conformation, irrespective of the ordering of the side groups. Recent ^{2}H NMR studies of LC side chain polymers, selectively deuterated at various positions of the spacer and polymer backbone, essentially confirm this model [132, 133]. Evidently, the microorder decreases abruptly from the last spacer segment to the polymer chain, which is only slightly distorted from its random coil conformation. The weak preferential orientation of the polymer backbone, either parallel or perpendicular to the director, was found to depend on the chain stiffness, in agreement with theoretical predictions [134–136]. However, recent small angle neutron scattering experiments are at variance with these observations [137], indicating that the preferential chain orientation is always perpendicular to the director [138–141]. Further studies are required to clarify this point.

The most prominent feature of LC side chain polymers is their ability to preserve long-range orientational order in the solid state. Inspection of Fig. 17 reveals that the order parameters S_{zz} are quantitatively retained when the polymers **1, 2** are cooled below the glass transition temperature. No change in S_{zz} is observed over a period of several years, keeping the samples at room temperature [35] Thus, in contrast to the low molecular weight analogues **5**, crystallization in the solid state is prohibited and long-range orientational order preserved. This particular behaviour of LC side chain polymers opens new applications in the storage technology, as we shall see in the next section.

4.3 Macroscopic Alignment

Macroscopic alignment of LC side chain polymers has been achieved using electric, magnetic and shear fields [27, 35, 142–146]. Generally, the degree of macroorder,

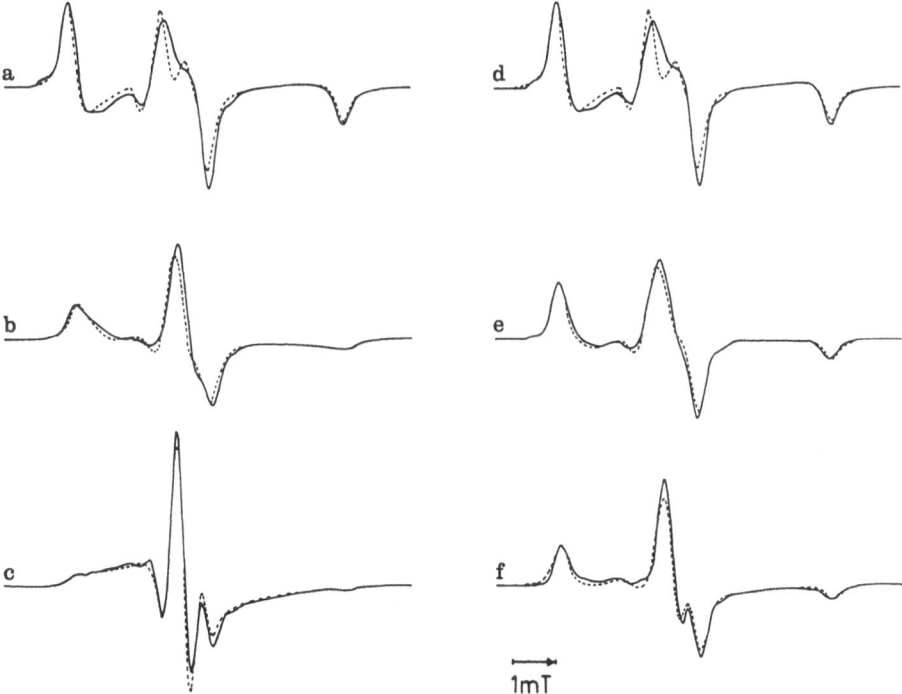

Fig. 18a–f. Experimental (———) and simulated (– – –) ESR spectra of the cholestane spin probes in macroscopically aligned samples of the side chain polymer **1** with a molecular weight of $\overline{M}_n = 4500$ (left column) and of $\overline{M}_n = 14000$ (right column). The spectra refer to T = 253 K and three different orientations ϱ of quartz plate normal and magnetic field, namely $\varrho = 0°$ **(a, d)**, $\varrho = 50°$ **(b, e)** and $\varrho = 90°$ **(c, f)**. Simulations were obtained with the parameters given in Table 2 and in Figs. 15 and 17. The macroorder parameters employed are $S_{x'x'} = 0.8$, $S_{z'z'} = -0.5$ (left column) and $S_{x'x'} = 0.2$, $S_{z'z'} = -0.5$ (right column)

obtained for a particular system, sensitively depends on the orientation method and the alignment conditions. In case of the side chain polymers 1, 2 and model compounds 5 macroscopic alignment was attempted shortly below the clearing temperature ($T^* = 0.98$) with the quartz plates of the electric field cell (50 kV/cm, 50 kHz) parallel to a variable magnetic field ($0.33\,T \leq B \leq 0.7\,T$).

In Fig. 18 the effect of the molecular weight on the macroorder is demonstrated. The ESR spectra refer to samples of polymer 1 ($\overline{M}_n = 4500$, left column; $\overline{M}_n = 14000$, right column) and three different orientations ϱ of quartz plate normal and magnetic field. Interestingly, at $\varrho = 0°$ both samples show identical lineshapes (top row). However, with increasing rotation angle significant differences appear (center row), which are most pronounced in the $\varrho = 90°$ spectra (bottom row). Evidently, the macroorder achieved depends, on the degree of polymerization [35].

Experimental values of the macroorder parameters $S_{x'x'}$, $S_{y'y'}$ and $S_{z'z'}$ (see Sect. 3.3), obtained for the various systems, are listed in Table 5. Drastic variations are observed. Apparently, a complete uniform alignment ($S_{x'x'} = 1.0$, $S_{y'y'} = S_{z'z'} = -0.5$) of the director axes is achieved for the LCP 2 and the monomeric analogue 5. The observed macroorder parameters are consistent with a positive diamagnetic ($\Delta\chi > 0$) and a negative dielectric anisotropy ($\Delta\varepsilon < 0$).

For the LCPs 1 (short spacer), the situation is more complicated. One sees that $S_{z'z'}$, still exhibits the limiting value of $S_{z'z'} = -0.5$, implying complete two-dimensional alignment with respect to the electric field ($\Delta\varepsilon < 0$). However, variation of $S_{x'x'}$ in the range $0.2 \leq S_{x'x'}, \leq 0.65$ indicates only partial ordering by the magnetic field. Note that $S_{x'x'}$ critically depends on the molecular weight of the system. For samples with $\overline{M}_n = 14000$, no magnetic field induced alignment is observed in contrast to samples with $\overline{M}_n = 4500$, where partial ordering occurs [35].

It has been shown that the magnetic field induced deformations of the macroorder only occur above a threshold field B_c, which is directly related to the effective elastic constant k of the system [147, 148]. Preliminary ESR studies [35] of the LCPs 1 (short spacer) yield elastic constants which exceed those of LMCLs [149] by one to two orders of magnitude. Moreover these elastic constants depend on the molecular weight of the side chain polymer. In case of a sufficient long spacer (LCPs 2), however, the threshold fields, and therefore the elastic constants are comparable to those of LMLCs [35], in agreement with other studies in electric and magnetic fields [27, 123].

Interestingly, the macroorder of LC side chain polymers can be frozen in at the glass transition. No electric or magnetic field is required to maintain the director distribution

Table 5. Parameters characterizing the macroscopic alignment of liquid crystal side chain polymers and low molecular weight analogues[a]

Macroorder parameter	LCP 1 $\overline{M}_n = 4500$	LCP 1 $\overline{M}_n = 14000$	LCP 2 $\overline{M}_n = 14000$	LMLC 5 $\overline{M}_n = 426$
$S_{x'x'}$	0.65	0.20	1.0	1.0
$S_{y'y'}$	−0.15	0.30	−0.50	−0.50
$S_{z'z'}$	−0.50	−0.50	−0.50	−0.50

[a] Macroscopic alignment was attempted at $T^* = 0.98$ with the quartz plates of the electric field cell (50 kV/cm, 50 kHz) parallel to a magnetic field of 0.33 T.

in the sample. Therefore, these polymers can be used as storage material. The information, which is inserted in the LC phase, can be stored permanently in the glassy state of the material. This exceptional property of LC side chain polymers has been applied in new thermo-optic storage devices [32, 150–154]. Further applications in the field of holographic data storage [155] and non-linear optics [156–158] are currently being explored.

Finally, macroscopically oriented fibres or films can be prepared by drawing of the side chain polymers [129, 130, 159, 160]. Generally, the polymer backbone aligns parallel to the drawing axis, whereas the arrangement of the side groups depends on the chemical structure of the systems [129, 159, 160]. In any case, however, the high degree of macroorder achieved by mechanical forces is quantitatively retained in the solid state of these systems. Potential applications in the field of integrated optics are recently being explored in relation to the development of LC elastomers [161, 162].

5 Liquid Crystal Main Chain Polymers

Molecular order and dynamics of the LC main chain polymers **3, 4** and corresponding model compounds **6–8** (see Fig. 13) have been studied by pulsed ^{2}H NMR [10, 63, 94–96]. The major results, presented in the following, reveal the prominent molecular features distinguishing LC main chain polymers from their monomeric analogues. Again, attempts are made to correlate the observed molecular and macroscopic properties of these systems.

5.1 Molecular Dynamics

^{2}H NMR relaxation studies, using 1D and 2D techniques, have been employed to evaluate the molecular dynamics of the systems. Representative results for polymer **4**, selectively deuteriated at different sites in the repeating unit (see Fig. 13), are depicted in Figs. 19–24. The observed ^{2}H NMR spectra and relaxation curves, varying drastically with the label position, demonstrate the power of the method.

Figure 19 shows quadrupole echo lineshapes of polymer **4** (α-CD$_2$) at five different temperatures [10]. The spectra refer to parallel orientation of alignment axis and magnetic field and characterize the various phases of the LC polymer. Except for an isotropic line in the high temperature spectrum (Fig. 19a) all lineshapes indicate highly ordered media. In addition, the spectra of the anisotropic melt (Figs. 19a, b) reflect fast inter- and intramolecular motions. Cooling below the melting point causes drastic spectral changes. The spectra of the solid state (Figs. 19c–e) now indicate two overlapping components, associated with mobile and immobile deuterons in distinct regions of the polymer.

They are easily separated by various relaxation experiments [10, 96]. Figure 20 shows partially relaxed ^{2}H NMR spectra of polymer **4** (α-CD$_2$) at four different pulse separations t_1 in a saturation recovery and a quadrupole echo sequence (see Fig. 6). The spectra refer to the same temperature and parallel orientation of alignment axis and magnetic field. Again two spectral components are observed. The central peaks refer to the mobile fraction of the polymer while the outer peaks correspond to the

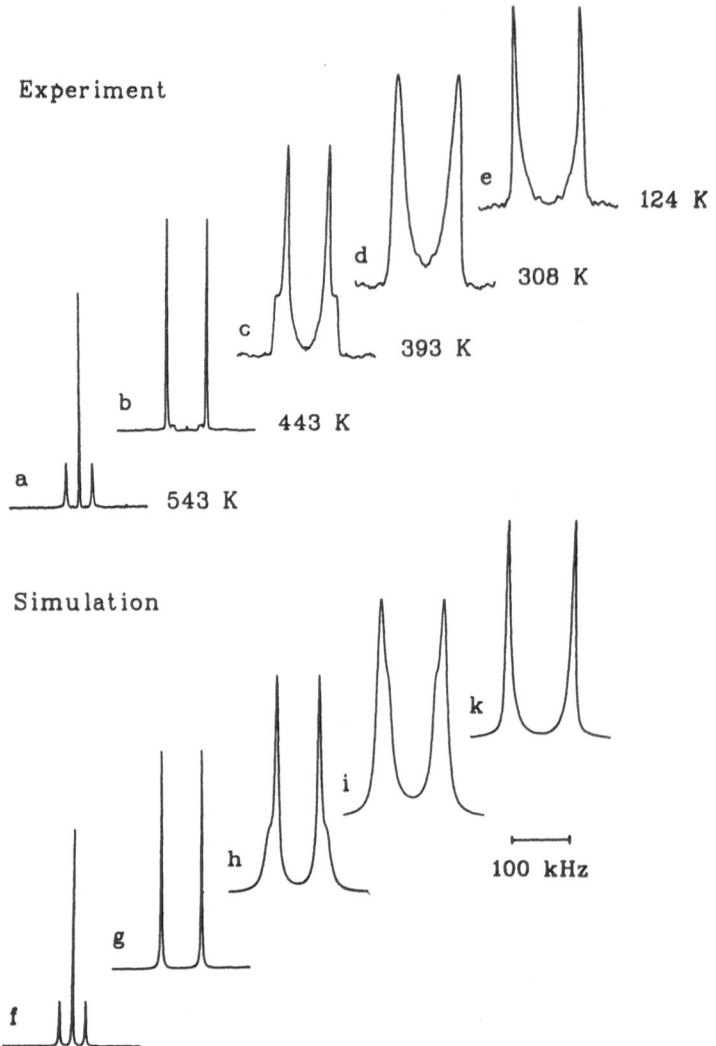

Fig. 19a–k. Experimental (upper row) and simulated (lower row) ^{2}H NMR spectra of main chain polymer **4** (α-CD$_2$), at five different temperatures. The spectra refer to quadrupole echo sequences ($t_1/2 = 20$ μs) and parallel orientation of alignment axis and magnetic field ($\varrho = 0°$). Simulations were obtained with the parameters given in Table 3 and in Figs. 25, 28 and 29. Mole fraction of isotropic component (19f) $x_i = 0.55$ and of immobile component (19h–19k) $x_{im} = 0.55$

immobile part. One recognizes that with increasing t_1 significant spectral changes occur, resulting from different spin lattice relaxation times T_{1Z} (top row) and spin-spin relaxation times T_{2E} (bottom row). Apparently, the mobile fraction relaxes somewhat faster than the immobile one.

In Fig. 21, the spin lattice relaxation times T_{1Z} (full and open squares) and the spin-spin relaxation times T_{2E} (full circles) are plotted as a function of $1/T$ [10, 96]. All relaxation times refer to polymer **4** (α-CD$_2$) and parallel orientation of alignment

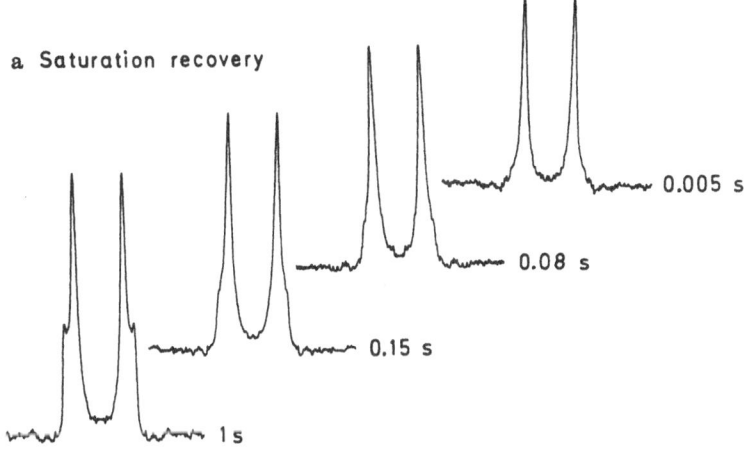

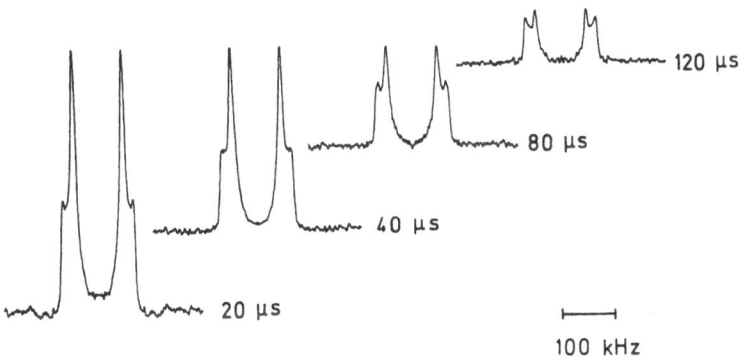

Fig. 20a,b. Experimental ^{2}H NMR spectra of main chain polymer **4** (α-CD$_2$) at T = 373 K and four different pulse separation times in a saturation recovery **a)** and a quadrupole echo sequence **b)**. The spectra refer to parallel orientation of alignment axis and magnetic field ($\varrho = 0°$)

axis and magnetic field. The values of T_{1Z} were obtained by recording the echo amplitude of an inversion recovery sequence (see Fig. 6) as a function of t_1 and by decomposing this function into two exponentials. One sees that the spin lattice relaxation times of the mobile (T_{1Z}^m) and immobile fraction (T_{1Z}^{im}) differ by a factor of 10–50. Interestingly, T_{1Z}^m passes through a sharp minimum shortly below the melting point, characteristic of a single motional process.

The spin-spin relaxation times T_{2E} were obtained from the decay of the echo amplitude in a qudrupole echo sequence (see Fig. 6). Since a *unique* decomposition into two exponentials was impossible the given T_{2E} data present bulk relaxation times and refer to both mobile and immobile deuterons. One sees that T_{2E} first decreases with decreasing temperature, then passes through a minimum and finally increases again.

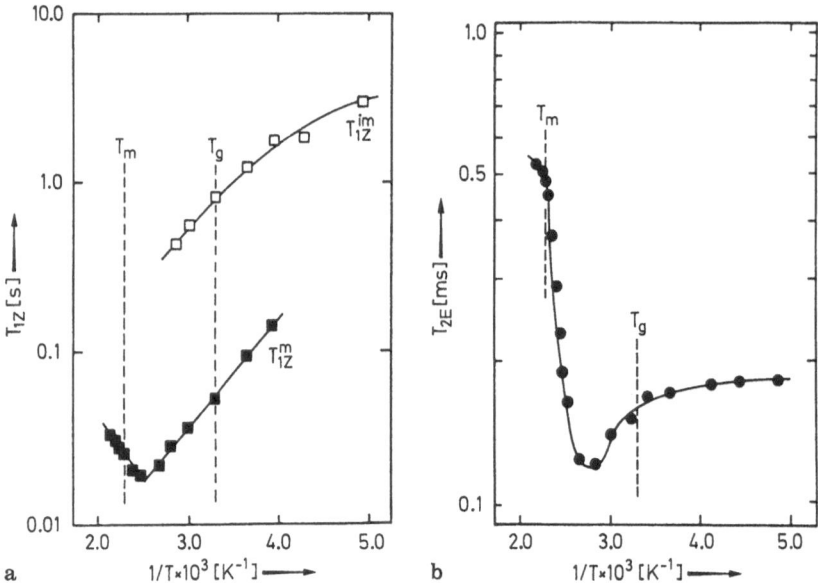

Fig. 21 a, b. Temperature dependence of deuteron spin relaxation times of main chain polymer **4** (α-CD$_2$) at parallel orientation of alignment axis and magnetic field ($\varrho = 0°$). **a)** Spin lattice relaxation times of mobile (T_{1Z}^m) and immobile component (T_{1Z}^{im}). **b)** Bulk spin-spin relaxation times T_{2E}, representing an average for mobile and immobile deuterons. *Dashed lines* indicate different phase transitions. The *solid lines* represent best fit simulations of the relaxation times, employing the relaxation model of Sect. 3.3 and the parameters of Figs. 25, 28 and 29

This striking behaviour, expected on theoretical grounds, is characteristic of motions in the MHz regime [10, 49].

It should be noted, however, that two spectral components can only be observed by *slowly* cooling or annealing of the samples. This is illustrated in Fig. 22, depicting ^{2}H NMR powder spectra of polymer **4**, deuterated at the outer phenyl ring (φ_0, see Fig. 13). Note that the quenched sample (left column) exhibits a complex spectral feature, associated with mobile deuterons. In contrast, the spectra of the annealed sample (right column) indicate two different components, referring to mobile and immobile deuterons in distinct regions of the polymer. Apparently, the phenyl rings of the latter fraction are completely rigid on the time scale of the NMR experiment, as can be deduced from the Pake diagram (broad spectral component), visible in all spectra.

In order to differentiate the various motional modes, occurring in LCPs (see Fig. 4), 2D relaxation experiments have been carried out [63, 64]. Representative examples are shown in Fig. 23. The 2D relaxation spectra (left column) and corresponding contour plots (right column) refer to annealed samples of LCP **4**, specifically deuteriated in the outer phenyl ring (Fig. 23a) and first spacer segment (Figs. 23b, c) respectively. As discussed earlier (see Sect. 3.1), the shape of the contour plot is highly indicative of the type of motion, responsible for spin relaxation. Apparently, the distinct local motions, namely ring flips (Fig. 23a) and tetrahedral jumps (Fig. 23b), can be differentiated from the contour plots of a quadrupole echo sequence. Similar results are obtained from inversion recovery sequences (Fig. 23c). Thus, by applying this 2D techniques to different pulse sequences, the motional modes can be differentiated over an extremely wide dynamic range, extending over ten orders of magnitude [63, 64].

A further application of this technique is shown in Fig. 24, depicting 2D relaxation spectra of LCP **4**, deuteriated at different positions along the aliphatic spacer, namely at the α-, β- and δ-segment. All spectra refer to quadrupole echo sequences and annealed samples. Inspection of Fig. 24 reveals a significant dependence of the chain

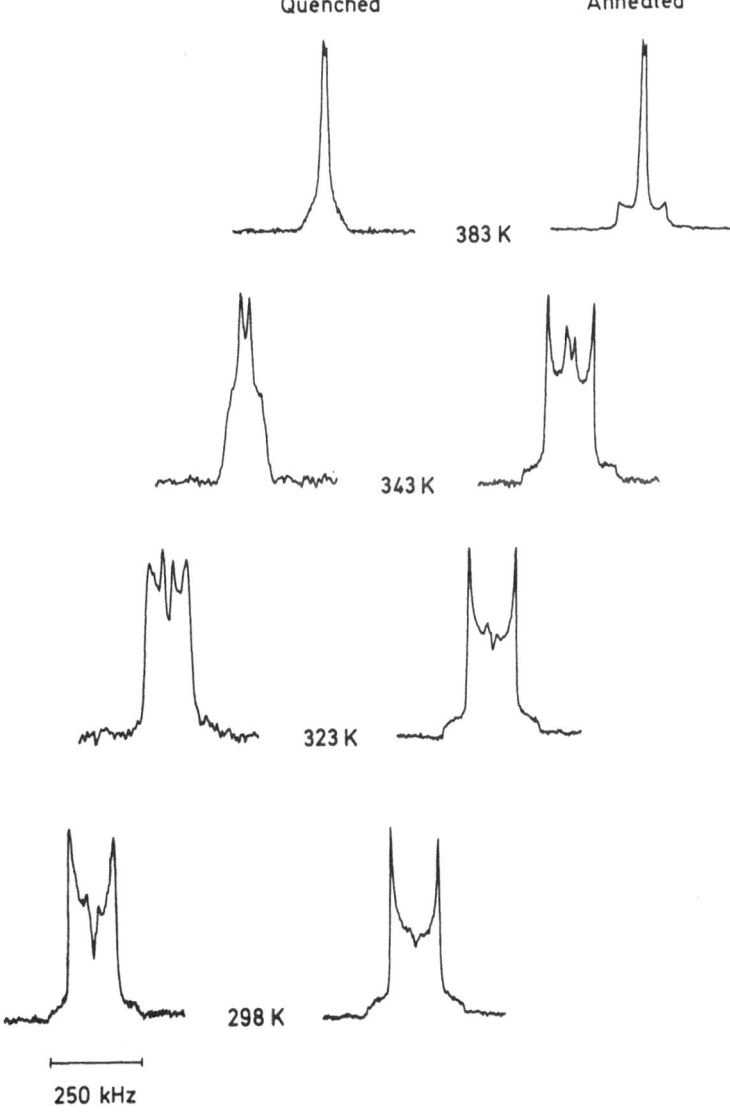

Fig. 22. Experimental ^{2}H NMR spectra of polydomain samples of main chain polymer **4**, deuteriated at the outer φ_a) phenyl rings. The spectra refer to a quenched (left row) and an annealed sample (right row) at four different temperatures

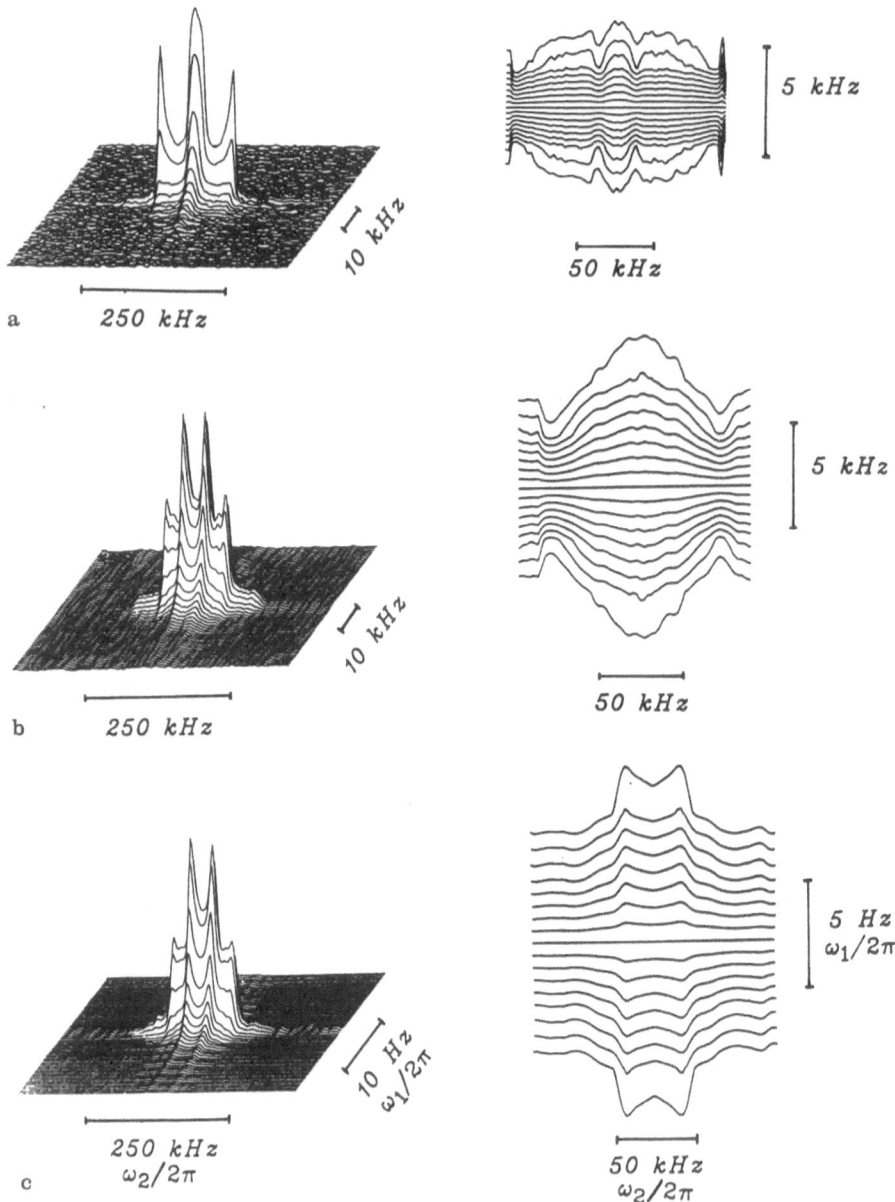

Fig. 23a–c. Experimental two-dimensional ^{2}H NMR relaxation spectra of an annealed sample of main chain polymer **4** at T = 363 K. **a)** Deuteriated at the outer phenyl rings (φ_0); quadrupole echo sequence. **b)** Deuteriated at the α-segment of the spacer; quadrupole echo sequence. **c)** Deuteriated at the α-segment of the spacer; inversion recovery sequence. Left side: Stack plots. Right side: Normalized contour plots (contours in units of 10% of maximum amplitude)

mobility on the label position, increasing from the α- to the δ-segment. Evidently, there is a mobility gradient along the spacer, clearly reflected by the contraction of the contour lines in the same direction. Interestingly, the mobility gradient is exhibited by both polymer fractions, appearing below the melting point.

Quantitative information about molecular dynamics in the various polymer phases has been obtained by computer simulations [10] of the ^{2}H NMR relaxation experiments. The lower row spectra in Fig. 19 and the solid lines in Fig. 21 represent best

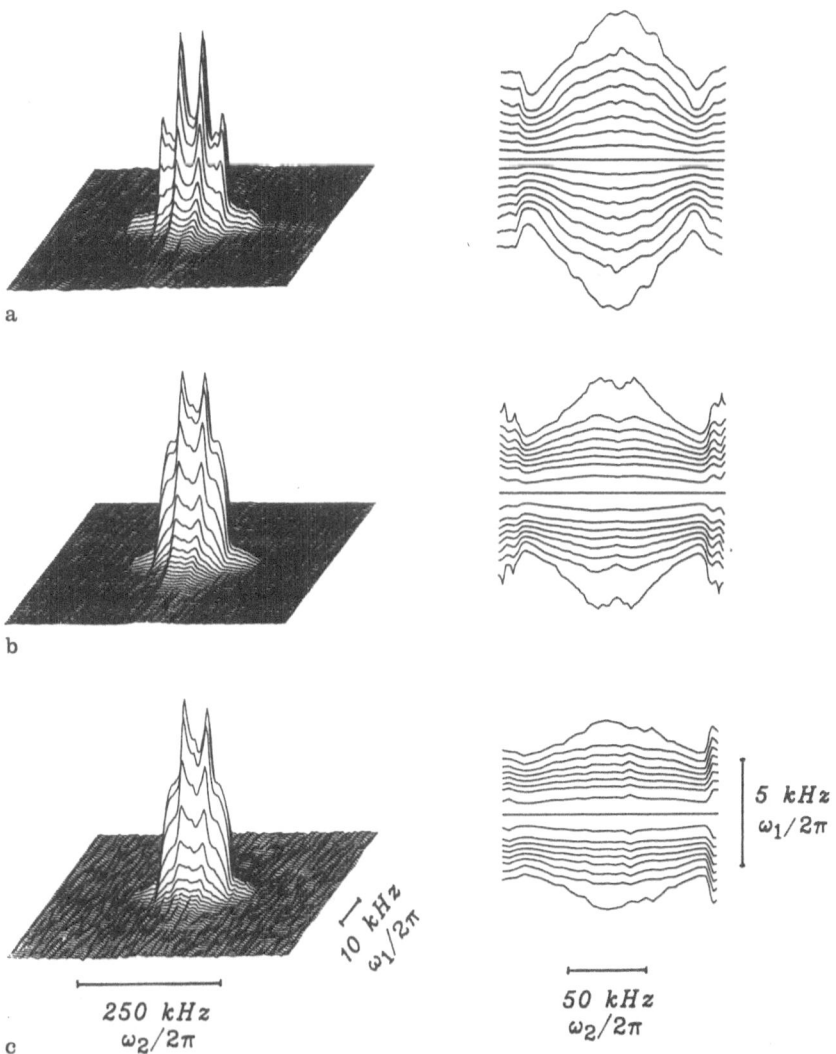

Fig. 24a–c. Experimental two-dimensional ^{2}H NMR relaxation spectra (quadrupole echo sequences) of an annealed sample of main chain polymer **4** at T = 363 K. **a)** Deuteriated at the α-segment of the spacer. **b)** Deuteriated at the β-segment of the spacer. **c)** Deuteriated at the δ-segment of the spacer. Left side: Stack plots. Right side: Normalized contour plots (contours in units of 10% of maximum amplitude)

fit simulations. In Fig. 25 the correlation times for the various motions of LCP **4** are plotted as a function of $1/T$. They refer to rotation about the long axis (full circles) and *trans-gauche* isomerization (full and open squares) of the first spacer segment (α-CD$_2$), respectively. For comparison, the correlation times for overall rotation of the low molecular weight analogues **7** (full triangles) and **8** (full diamonds) are depicted, likewise [95]. One sees that the Arrhenius plots are generally linear, and activation energies for the various motions have been determined from the slopes of the straight lines. They are listed in Table 6, together with the anisotropy ratios $\tau_{R\perp}/\tau_{R\|}$.

In the nematic range close to the melting point the correlation times for *trans-gauche* isomerization of LCP **4** are $\tau_J \sim 10^{-9}$ s, which is two orders of magnitude slower than in the model compounds **7** and **8** [95]. The average activation energy for this process of $E_J \sim 16$ kJ/mol (see Table 6) can be compared with the potential barrier height encountered in rotational isomerization about C—C bonds [91]. As mentioned above, the *trans-gauche* isomerization rate depends on the label position, increasing from the α- to the δ-segment. Evidently, there is a mobility gradient along the alkyl chains. In the *pendent groups* of the monomers **8** τ_J decreases by two orders of magnitude from the α- to the δ-position [95], in agreement with theoretical predictions [163]. On the other hand, a much less pronounced mobility gradient is observed

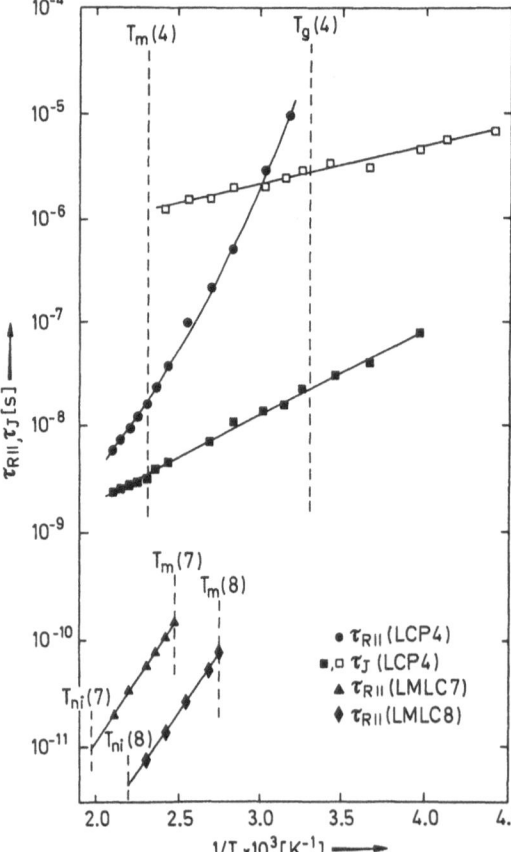

Fig. 25. Arrhenius representation of various correlation times, characterizing the molecular dynamics of main chain polymer **4** and low molecular weight analogues **7** and **8**. *Circles* (chain rotation) and *squares* (trans-gauche isomerization at α-segment) refer to polymers, *diamonds* (chain rotation) correspond to the dimers, and *triangles* (chain rotation) denote the monomers. *Dashed lines* indicate different phase transitions. Mobile and immobile components, observed for the polymers below the melting point, are distinguished by *full* and *open symbols*

Table 6. Dynamic parameters of liquid crystal main chain polymers and low molecular weight analogues

Liq. cryst. system	Phase[a]	Rot. activation energies		Anisotropy ratios
		$E_{R\parallel}$/(kJ/mol)	E_J[b]/(kJ/mol)	$\tau_{R\perp}/\tau_{R\parallel}$
LCP 4	n	48.3	15.9	11
LCP 4	bp(T > T_g)	48.3–71	15.9 (6.7)[c]	11–34
LCP 4	bp(T < T_g)	–	15.9 (6.7)[c]	–
LMLC 7	n	46.1	16.3	25
LMLC 8	n	39.9	16.1	8

[a] n = nematic, bp = biphasic region, T_g = glass transition temperature.
[b] *Trans-gauche* isomerization at α-segment.
[c] Crystalline component.

for the *spacers* of the polymers, presumably because of the increased conformational order in these systems (see Sect. 5.2).

The correlation times 10^{-9} s $< \tau_{R\parallel} < 10^{-8}$ s for overall rotation of the LCPs **4** in the nematic phase are at least two orders of magnitude slower than those observed for the model compounds **7** and **8**, in agreement with previous ESR studies [164]. This is reflected in the bulk viscosities of the polymers, exceeding those of the momomers by nearly a factor of 1000 [165]. The correlation times of the model compounds of 7×10^{-12} s $< \tau_{R\parallel} < 1.5 \times 10^{-10}$ s correspond to values reported for other LMLCs [73, 166–168]. In addition, the evaluated anisotropy ratios of $\tau_{R\perp}/\tau_{R\parallel} = 25$ (dimer **7**) and $\tau_{R\perp}/\tau_{R\parallel} = 8$ (monomer **8**) are in excellent agreement with the theoretical values of 26.5 and 8.7, calculated from the corresponding molecular dimensions [169].

The fact that a consistent description of the overall molecular motion can be obtained for all positions of labelling in the polymers **4** and model compounds **7, 8** supports the motional model (see Sect. 3.3). It further demonstrates the ability to discriminate between the different motional modes by simulation of 1D and 2D relaxation experiments [10, 64]. The temperature dependence yields activation energies of 39.9 kJ/mol $< E_{R\parallel} < 48.3$ kJ/mol (see Table 6), which clearly distinguish the intramolecular from the intermolecular motion. The segmental motion has activation energies that are comparable to the barrier height for *trans-gauche* isomerism [91], whereas the activation energies for overall motion are two to three times greater.

Molecular motions in LCs may occur as isolated or collective modes (see Fig. 4). For the latter mechanism, known as order director fluctuations, a continuous distribution of correlation times is expected [36, 37, 170–173]. Recent proton T_{1Z} dispersion measurements of the LCPs **4** and corresponding LMLCs **7** and **8**, carried out over a frequency range of five orders of magnitude (10^3 Hz $< \omega_0/2\pi < 3 \times 10^8$ Hz), clearly show that collective order fluctuations contribute to the relaxation process only at extremely low frequencies in the kHz regime, whereas the conventional MHz range is dominated by reorientations of individual molecules [174].

For nematic LCs, theory predicts a characteristic dispersion law $T_{1Z}(\omega_0) \propto \omega_0^{1/2}$ [36, 37]. This is exactly what we observe for the monomeric **8** and dimeric systems **7**. Although a somewhat higher exponent is evaluated for the polymers **4**, there is no doubt that collective order fluctuations occur in these systems, likewise [174].

Below the melting point the situation is more complicated. As mentioned above the dynamics now depends on the thermal history. Quenched samples exhibit relatively fast chain motions comparable to those in the nematic melt. For annealed samples two components are observed, which we assign to a LC and a crystalline phase. Decomposition of various relaxation curves into two components yields a crystallinity of 55% ± 5, practically independent of temperature [10, 96, 175]. A similar heterogeneity is observed for ordinary polymers, which exhibit amorphous and crystalline phases [176]. Accordingly, melt and glass transitions are detected in the DSC thermograms of both types of polymers [96, 177].

Let us now discuss the dynamics of annealed samples in more detail. Figure 25 shows a drastic motional decrease for the crystalline component (open symbols) at the melting point. Apparently, all intermolecular motions abruptly cease below T_m. Similar is true for the local motions of the phenyl rings (see Fig. 22). Thus, only slow *trans-gauche* isomerization of the spacer (10^{-5} s $< \tau_J < 10^{-6}$ s) can be detected in the crystalline state. The low activation energy of $E_J = 6.7$ kJ/mol for this jump process agrees with previous T_{1z} dispersion measurements on solid paraffins [178]. Apparently, there is a small but significant flexibility gradient along the spacer with τ_J decreasing from the α- to the δ-position (see Fig. 24).

In contrast, the dynamics of the LC component (full symbols) continues into the biphasic region (crystalline and LC phase) without any significant change at the melt transition [10]. Note, however, that the Arrhenius plot for chain rotation and chain fluctuation is not linear, the apparent activation energy increasing with decreasing temperature [10]. Thus all intermolecular motions gradually freeze (*glass transition process* [6, 179, 180]) and at temperatures T $< T_g$ intramolecular motions are the dominant process [10, 175, 181]. In fact, we have been able to detect *trans-gauche* isomerization even at T $= 130$ K with a correlation time of $\tau_J \sim 10^{-4}$ s [10]. Similarly, ring flips could be followed down to temperatures 100 K below the glass transition [182].

Figure 25 clearly shows the coexistence of LC and crystalline components, differing drastically in their molecular dynamics. In that respect thermotropic main chain polymers [183, 184] resemble ordinary polymers, which exhibit amorphous and crystalline phases [40–42, 176, 185–187]. However, a broad distribution of correlation times is generally evaluated in all but the lowest molecular weight systems. In contrast, molecular reorientation in the LCPs 4 *above* T_g appears to occur essentially by single processes, where any distribution of correlation times must be restricted to less than one decade. This is particularly obvious from Fig. 21, which depicts a sharp T_{fz} minimum, characteristic of a single process. It should be noted, however, that a distribution of correlation times of 2.5 decades is evaluated for the ring flip motions *below* T_g [182].

Comparable investigations of the molecular dynamics in LC main chain polymers are rare [6, 188–192]. ESR studies, using nitroxide spin probes, have been employed to evaluate the overall dynamics of these systems [188, 190]. The results are in qualitative agreement with the present findings. A complimentary ^{2}H NMR study of combined main-chain/side-chain LCPs has recently appeared [114]. Interestingly, the rotational rates reported for intermolecular motions are very similar to those exhibited by "pure" main chain systems [114, 115]. Evidently, the laterally attached side group mesogens do not provide a significant rotational hindrance. Moreover, as described above, the apparent activation energies for chain rotation and chain fluctuation increase with decreasing temperature, leading to a gradual freeze of all intermolecular motions (*glass transition process*). Below T_g intramolecular motions (ring flips) are reported to be the dominant process [114, 115].

5.2 Molecular Order

Our present knowledge about the molecular order of LC main chain polymers and corresponding model compounds originates to a large extent from dynamic NMR studies, using ^2H nuclei [10, 94–96, 175, 181]. These investigations provide detailed information about the molecular organization, comprising orientational and conformational order in the various mesophases. For a reliable molecular characterization, however, the knowledge of the molecular geometry is required. Angular de-

a Experiment

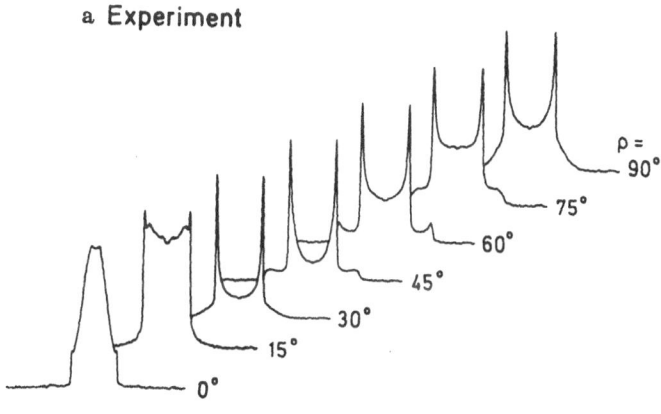

b Calculation

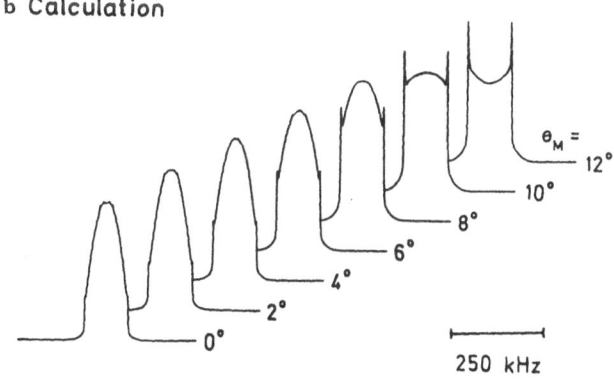

250 kHz

Fig. 26a. Experimental ^2H NMR spectra of macroscopically aligned main chain polymer **4**, deuteriated at the central phenyl ring (φ_C), at seven different orientations ϱ of alignment axis and magnetic field. The spectra refer to quadrupole echo sequences ($t_1/2 = 25$ µs) and $T = 253$ K. **b)** Calculated ^2H NMR spectra, showing the dependence of the molecular tilt angle θ_M. The spectra were calculated with $\varrho = 0°$, $\tau_j \geq 10^{-3}$ s and the parameters given in Figs. 25 and 29

pendent lineshapes offer the chance to evaluate the various molecular tilt angles unambiguously.

Figure 26a depicts the angular variation of quadrupole echo spectra at T = 253 K. The lineshapes refer to LCP **4**, deuteriated in the central phenyl ring (φ_c, see Fig. 13), and seven different orientations ϱ of alignment axis and magnetic field. Very large spectral changes are observed when the sample is rotated, showing that LC order is maintained in the solid and glassy state of the polymer. The pronounced angular dependence of the spectra was employed to evaluate the molecular geometry of the systems. Figure 26b shows calculated lineshapes for $\varrho = 0°$ and different tilt angles θ_M, characterizing the orientation of the order tensor relative to the central phenyl ring (see Fig. 12). Apparently, the tilt angle is close to zero in this case.

Table 3 summarizes the geometry parameters, used in the analysis of the ^{2}H NMR experiments. One sees that the model compounds **7** and **8** exhibit the same molecular geometry as the polymers **4** [95]. Generally, molecular tilt angles from angular dependent spectra are in good agreement with those evaluated from motionally averaged spectra at a single orientation, employing inequivalent deuterons.

Figure 27 compares fast-rotational ^{2}H NMR spectra of polymers (LCP **4**, Fig. 27a–c), dimers (LMLC **7**, Fig. 27d–f) and monomers (LMLC **8**, Fig. 27 g–i), demonstrating the effect of the molecular weight on the degree of order. The lineshapes refer

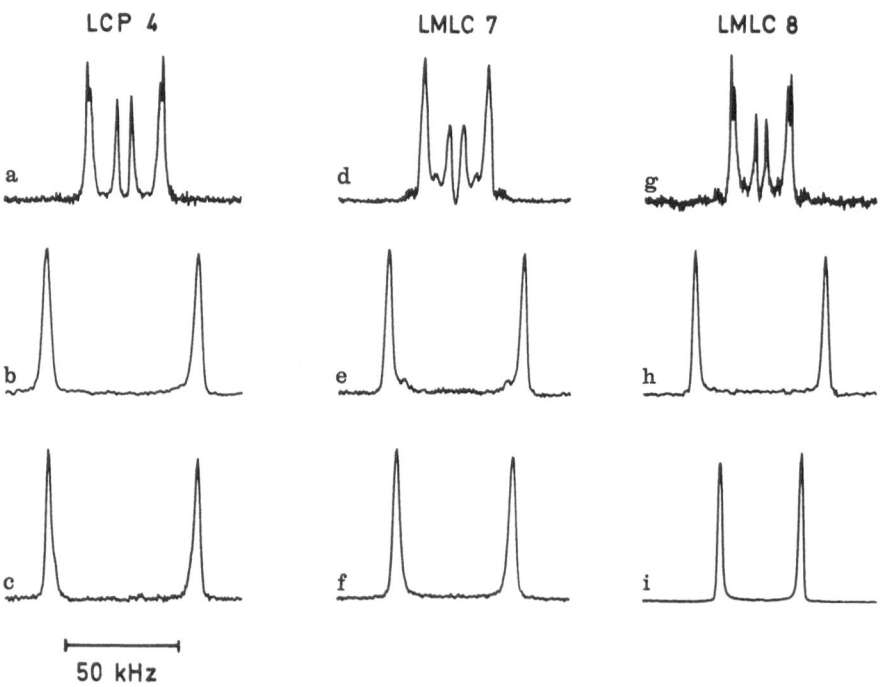

Fig. 27a–i. Experimental ^{2}H NMR spectra of main chain polymer **4** and low molecular weight analogues **7** and **8** in the nematic phase. The spectra are recorded for the same reduced temperature (T* = 0.88) and parallel orientation of alignment axis and magnetic field ($\varrho = 0°$). **a)** Polymer **4** (φ_C-D$_3$), **b)** polymer **4** (α-CD$_2$), **c)** polymer **4** (δ-CD$_2$), **d)** dimer **7** (φ_C-D$_3$), **e)** dimer **7** (α-CD$_2$), **f)** dimer **7** (δ-CD$_2$), **g)** monomer **8** (φ_C-D$_3$), **h)** monomer **8** (α-CD$_2$), **i)** monomer **8** (δ-CD$_2$)

to the same reduced temperature ($T/T_{ni} = 0.88$) and parallel orientation of alignment axis and magnetic field. Note that systems labelled at the central phenyl ring (top row spectra) exhibit three different quadrupole splittings, owing to the presence of inequivalent deuterons [94, 95]. The inequivalence is due to different orientations of the various C–D bonds relative to the para-axis of the phenyl ring (see Table 3). In contrast, the NMR spectra of systems, deuteriated at the α- (center row spectra) and δ-segment (bottom row spectra) of the alkyl groups, show a single splitting. A comparison of the spectra of polymers **4**, dimers **7** and monomers **8** reveals a consistent smaller splitting for the latter compounds [95]. Evidently, molecular order also depends on the degree of polymerization.

The observed lineshapes were analyzed, employing the NMR model outlined above (see Sect. 3.3). An iterative fit of several angular dependent experiments for any given temperature provided detailed information about the molecular organization of the systems, comprising the conformational order of the alkyl chains and the orientational order of the mesogenic units. Conformational order is conveniently described in terms of segmental order parameters $S_{Z'Z'}$, and $S_{X'X'} - S_{Y'Y'}$, defined elsewhere [49]. Within the limits of a completely disordered segment, all four tetrahedron sites are equally populated, resulting in an order parameter $S_{Z'Z'} = 0$. At the other extreme, a fully extended chain is fixed to its all-*trans* conformation, and the order parameter becomes equal to unity.

In Fig. 28 these order parameters are plotted as functions of the reduced temperature $T^* = T/T_{ni}$. They refer to the monomers **8** (triangles), dimers **7** (squares) and polymers **4** (circles), respectively. The α- and δ-segments are distinguished by open and full symbols. At the isotropic-nematic transition ($T^* = 1.0$) order parameters with values between $(S_{Z'Z'})_{ni} = 0.61$ and $(S_{Z'Z'})_{ni} = 0.25$ are observed. One sees that conformational order of the α-segment is high. The observed minor differences be-

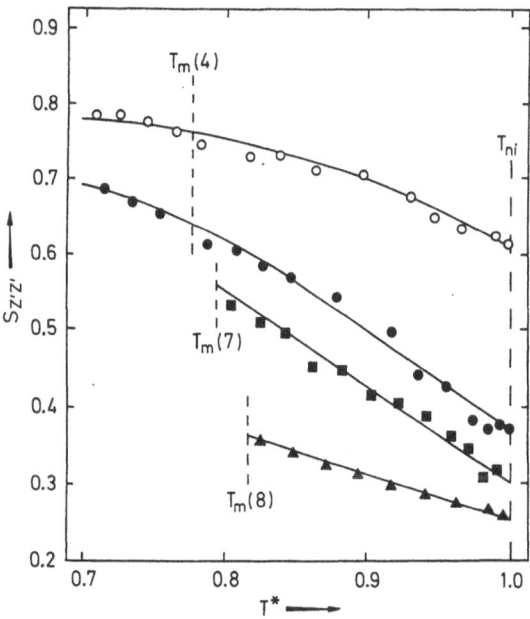

Fig. 28. Temperature dependence of conformational order of main chain polymers and low molecular weight analogues, expressed in terms of segmental order parameters $S_{Z'Z}$ at various chain segments. *Open symbols* refer to the α-positions, while *full symbols* denote the δ-positions. The order parameters refer to main chain polymer **4** (○, ●), dimer **7** (■) and monomer **8** (▲). *Dashed lines* indicate different phase transitions. $T^* =$ reduced temperature

tween monomer, dimer and polymer at T_{ni} (not shown) [95] completely vanish at the melt transition. The high $S_{Z'Z'}$ values for the α-segments can be rationalized by the close vicinity of the mesogenic cores.

For the δ-segments the dependence of the conformational order on the specific molecular structure is much more pronounced. At the clearing point T_{ni} we observe a stepwise increase in $S_{Z'Z'}$ from $(S_{Z'Z'})_{ni} = 0.25$ (monomers) via $(S_{Z'Z'})_{ni} = 0.30$ (dimers) to $(S_{Z'Z'})_{ni} = 0.37$ (polymers). Note, however, that all $S_{Z'Z'}$ values of the δ-position are significantly smaller than those evaluated for the α-segments. Apparently, there is an order or "flexibility" gradient along the chains, which varies from system to system studied.

Table 7 summarizes the results for the LC main chain polymers and corresponding model compounds. Note that the populations $P'_{eq}(\Omega_m) \equiv n_1, \ldots n_4$ of all conformational states [91], accessible at a particular segment, are presented. Apparently, the segmental order parameter $S_{Z'Z'}$, evaluated from the populations n_m, significantly decreases from the α- to the δ-positions. Interestingly, the anisotropy of the segmental order $S_{X'X'} - S_{Y'Y'}$ remains constant for the various positions. Interpretation of these chain ordering data in terms of statistical mechanical models of LCs presents a challenging theoretical problem [193–195].

Inspection of Table 7 reveals that the dimers 7 are good model compounds for the polymers 4. This result can be rationalized by the presence of *two* mesogenic units forcing the flexible spacer in a more extended configuration. Similarly, the increase in orientational order from the monomers 8 to the dimers 7 can be understood by intramolecular order transfer via the extended spacer.

Decreasing the temperature increases conformational order at all positions (see Fig. 28). In the case of the monomers, however, the increase at the δ-segment is smaller than the corresponding increase of the α-segment, leading to an enhanced order gradient at lower temperatures [95]. In contrast, the segmental order parameters of the dimers and polymers converge upon cooling, resulting in a reduced order gradient [95, 196]. Thus, ultimately, a uniform segmental order for all spacer segments is expected and is actually observed for the LPCs in the solid state [10, 175, 181].

Note, however, that there are two different phases coexisting in these polymers below T_m (see Sect. 5.1). Only the LC component exhibits the segmental order, depicted in Fig. 28. For the crystalline fraction a temperature independent order para-

Table 7. Conformational order of liquid crystal main chain polymers and low molecular weight analogues, characterizing the order gradient in the nematic phase ($T^* = 0.85$)

Liq. cryst. system	Chain segment	Populations of the conformational states				Segmental order parameters		Orientational order parameter
		n_1	n_2	n_3	n_4	$S_{Z'Z'}$	$S_{X'X'} - S_{Y'Y'}$	S_{ZZ}
LCP 4	α-C	0.78	0.11	0.11	0.00	0.71	0.15	0.85
LCP 4	δ-C	0.68	0.14	0.14	0.04	0.58	0.14	0.85
LMLC 7	α-C	0.78	0.11	0.11	0.00	0.71	0.15	0.76
LMLC 7	δ-C	0.62	0.17	0.17	0.04	0.50	0.18	0.76
LMLC 8	α-C	0.78	0.10	0.10	0.02	0.71	0.11	0.71
LMLC	δ-C	0.50	0.20	0.20	0.10	0.34	0.14	0.71

meter of $0.8 < S_{Z'Z'} < 0.85$ is observed [10, 96], far below the limiting value of $S_{Z'Z'} = 1.0$ expected for a regular crystal. This finding corresponds to the minor enthalpy of fusion [96, 197], being considerably smaller than that of conventional semicrystalline polymers [176, 198]. Apparently, the regularity of the crystalline region [183, 184] is reduced by conformational disorder of the spacer [10, 175], in agreement with the high mobility described in the previous section.

On the other hand, the limiting segmental order parameter of $0.75 < S_{Z'Z'} < 0.80$, observed for the LC phase of polymer **4** [10, 175, 181], implies rather high *trans* populations throughout the spacer. Evidently, highly extended conformers prevail in the LC state of these systems. This finding is the most prominent feature distinguishing main chain polymers from side chain polymers [35, 108] and their low molecular weight analogues [193, 194, 199]. Predictions of statistical mechanical theories [195, 200–202] are in qualitative agreement with our NMR results. It appears that a number of unique properties, exhibited by these polymers, can be attributed to this high degree of conformational order [201].

We now discuss the orientational order of the LCPs **3**, **4** and low molecular weight analogues **6–8** in terms of the familiar oder parameters S_{ZZ} and $S_{XX} - S_{YY}$ [34]. They express the ordering of the molecular axis Z and the anisotropy of the orientational order, respectively. Angular dependent lineshapes of the polymers indicate that each repeating unit can be characterized by an order tensor axially symmetric along Z [10, 95]. Thus, within experimental error $S_{XX} - S_{YY} = 0$ for these systems, in contrast to similar combined LCPs, for which a small molecular biaxiality has been observed [114, 115].

In Fig. 29a the order parameters S_{ZZ} of the "even-numbered" systems LCP **4**, LMLC **7** and LMLC **8**, having ten methylene segments in the repeating unit (see Fig. 13), are plotted as functions of the reduced temperature T*. Triangles refer to monomers, squares denote dimers, while circles correspond to polymers. One sees that at T* = 1.0 all order parameters jump to finite values, showing a stepwise increase from $(S_{ZZ})_{ni} = 0.43$ (monomers) to $(S_{ZZ})_{ni} = 0.50$ (dimers) and $(S_{ZZ})_{ni} = 0.64$ (polymers) [95]. This dependence of orientational order on the degree of polymerization is reflected by the transition entropies ΔS_{ni} (see Table 1), which vary between $\Delta S_{ni} = 4.6$ J/(molK), $\Delta S_{ni} = 8.0$ J/(molK) and $\Delta S_{ni} = 12.1$ J/(molK). Similar thermodynamic data have been reported for other model compounds and parent LCPs [203–207].

Lowering the temperature increases S_{ZZ} for all nematogens to limiting values of $S_{ZZ} = 0.75$ (monomers), $S_{ZZ} = 0.80$ (dimers) and $S_{ZZ} = 0.90$ (polymers) at the corresponding melt transitions. Note that the orientational order of the main chain polymers **4** considerably exceeds that exhibited by side chain polymers [35, 108] and low molecular weight analogues [193, 194, 199]. As predicted [200, 201, 208–210], the polymer chains are highly ordered on a molecular level, in agreement with ESR [188] and proton NMR studies [211–213] on similar systems. This finding is further corroborated by recent small angle neutron scattering experiments, indicating *high* anisotropy ratios for the radii of gyration in these semiflexible polymers [214].

Within experimental error, the S_{ZZ} values of the dimers **7** correspond to the mean order parameters of the monomers **8** and polymers **4**, consistent with theoretical predictions [201] and experimental observations on other systems [215–217]. The pronounced increase in molecular order from the monomers to the dimers is apparently caused by intramolecular order transfer via highly extended spacers. Consequently,

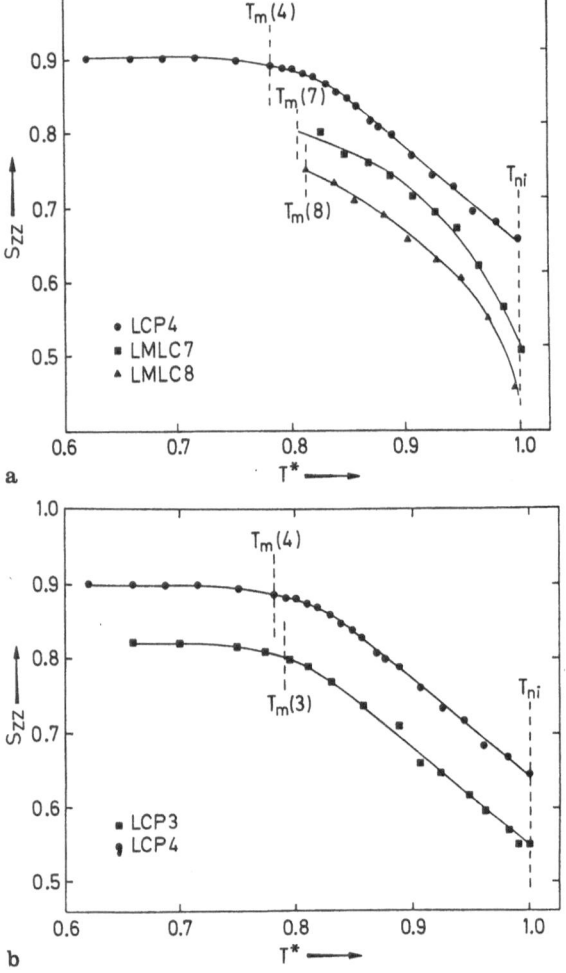

Fig. 29a, b. Temperature dependence of the orientational order of main chain polymers and low molecular weight analogues, expressed by the conventional order parameter S_{zz}. **a)** The order parameters refer to main chain polymer **4** (●), dimer **7** (■) and monomer **8** (▲). **b)** The order parameters refer to "odd-numbered" polymer **3** (■) and "even-numbered" polymer **4** (●). *Dashed lines* indicate different phase transitions. T^* = reduced temperature

linking of additional mesogenic units by flexible spacers should further increase the orientational order of these systems. Recent investigations on trimeric and tetrameric model compounds support these considerations [218, 219].

So far, the discussion has referred to systems having ten segments in the repeating unit. Reducing the alkyl chain length to nine methylene groups ("odd-numbered" systems) causes a pronounced change in the order parameter S_{zz}. This is illustrated in Fig. 29b, depicting the effect of the spacer parity on the orientational order of the polymers. Evidently, the values for the "odd-numbered" polymers **3** (squares) are about 10% smaller than those of the "even-numbered" systems **4** (circles). A similar result has been reported for the dimers **6** and **7** [95], in agreement with the corresponding clearing entropies of $\Delta S_{ni} = 6.1$ J/(molK) and $\Delta S_{ni} = 8.0$ J/(molK) (see Table 1). The pronounced even-odd effect also observed for other dimeric [215, 216] and polymeric LCs [211, 220–224] presents an interesting theoretical problem. Statistical mechanical treatments of this phenomenon have recently appeared [200,

201]. They are in substantial agreement with the present NMR results [10, 95, 225]. It should be noted, however, that the segmental order at the δ-position of even and odd numbered polymers (**3** and **4**) is essentially the same [225].

Interestingly, in the systems with ten spacer segments, the order parameter is retained quantitatively when the polymer is cooled below the melting point and glass transition [10, 175, 181, 226, 227]. No change in S_{zz} is observed after keeping the sample at room temperature. Thus, in contrast to conventional mesogens the long-range orientational order is preserved even upon crystallization [10]. This opens new applications in the field of high modulus fibers, as we shall see in the next section. In all systems studied $\overline{S_{zz}}$ is independent of the molecular weight of the polymers within the range $5000 < \overline{M}_n < 30000$, according to a plateau effect, commonly observed for LC main [220, 228] and side chain polymers [103, 104].

5.3 Macroscopic Alignment

Orientational order of LCs in bulk samples comprises the orientational order of the molecules with respect to the local director (microorder) and the macroscopic alignment of the director axes in the laboratory frame (macroorder). Various methods have been used to measure order parameters in LC main chain polymers, including the use of diamagnetic susceptibilities [217]. X-ray diffraction [229, 230], and IR dichroism [231], but these methods are essentially unable to separate the two types of orientational order. As mentioned earlier, however, dynamic magnetic resonance, including CW ESR and pulsed ^{2}H NMR, does not suffer from this deficiency.

Values of macroorder parameters $S_{z'z'}$ (see Sect. 3.3), obtained for the LCPs **4** and model compounds **7** and **8**, using various orientation methods, are listed in Table 8. For a random distribution of director axes $S_{z'z'} = 0$. In case of a complete alignment we expect $S_{z'z'} = 1.0$ or $S_{z'z'} = -0.5$, depending on the sign of the diamagnetic or dielectric anisotropy of the system.

Evidently, the degree of macroorder achieved depends not only on the orientation

Table 8. Parameters characterizing the macroscopic alignment of liquid crystal main chain polymers and low molecular weight analogues

Liq. cryst. system	Orientation method	Alignment condition	Reduced temp. T*	Macroorder parameters $S_{z'z'}$
LCP **4**	electric field[a]	E = 48 kV/cm	0.82	−0.5
LCP **4**	magnetic field	B = 7.0 T	0.82	1.0
LCP **4**	extrusion[b]	R_N = 49	0.67	0.9
LCP **4**	melt-spinning[c]	V_t/V_0 = 50	0.81	> 0.95
LMLC **7**	electric field[a]	E = 1.85 KV/cm	0.85	−0.5
LMLC **7**	magnetic field	B = 0.70 T	0.85	1.0
LMLC **8**	electric field[a]	E = 1.2 kV/cm	0.85	−0.5
LMLC **8**	magnetic field	B = 0.33 T	0.85	1.0

[a] Frequency ν = 50 kHz.
[b] R_N = draw ratio.
[c] V_t/V_0 = spin-draw ratio.

method but also on the alignment conditions. Since the dielectric anisotropy of all systems is negative ($\Delta\varepsilon < 0$), only a two-dimensional distribution of director axes is achieved using high frequency ($\nu = 50$ kHz) electric fields. The observed macroorder parameters of $S_{z'z'} = -0.5$, evaluated by dynamic ESR, [165, 188], indicate complete alignment with respect to the electric field. However, the required field strengths significantly differ, varying between 1.2 kV/cm $< E <$ 1.85 kV/cm for the model compounds 7, 8 and $E = 48$ kV/cm for the parent LCPs 4 [165]. Moreover, compared to conventional LCs [146, 147], saturation of the macroorder in the polymers 4 is slow, in agreement with observations on other main chain systems [232–236].

A uniform alignment of the director axes ($S_{z'z'} = 1.0$) is obtained by strong magnetic fields ($\Delta\varkappa > 0$). However, the required field strengths B depend critically on the degree of polymerization. For the monomers 8 complete alignment is achieved at $B = 0.33$ T, whereas for the dimers a field of $B = 0.7$ T is required [95, 165]. The corresponding field strengths of the polymers 4 exceed these values by a considerable amount. As reported previously [10], no alignment is achieved for $B = 1.5$ T, whereas complete macroscopic orientation iṣ achieved at $B = 7$ T. Similar results were obtained for other LC main chain polymers [229, 237–239]. It has been shown, that magnetic field induced director reorientation occurs only above a threshold field B_c, which is directly related to the elastic constant k of the system [147, 148]. Preliminary studies of the LCPs 4 yield elastic constants which exceed those of the corresponding model compounds 7 and 8 by two orders of magnitude [10, 165] in agreement with recent studies on other main chain polymers [232–236].

In addition, surface forces [240] and shear fields [96, 188, 241–243] have successfully been employed in orienting the LCPs. Solid state extrusion [10, 188] and melt-spinning [96] produce fibers, with nearly perfect alignment of the director axes ($S_{z'z'} \geqq$ 0.9). This is demonstrated in Fig. 30. The ^{2}H NMR spectra (top row) refer to melt-spun fibers of LCP 4 (α-CD_2) and five different orientations of fiber axis and magnetic field. Drastical lineshape changes are observed when the sample is rotated. A detailed analysis, based on spectral simulations (bottom row), provides the parameters of micro- and macroorder, summarized in Table 9 [96].

Generally, as-spun and annealed fibers exhibit the same high degree of conformational ($S_{z'z'} = 0.80$) and orientational order ($S_{ZZ} = 0.90$) of the polymer chains. In addition, practically all director axes are aligned in draw direction, evidenced by a macroorder parameter of $S_{z'z'} = 0.90$. Note, however, that this value still increases with decreasing spinning temperature to $S_{z'z'} > 0.95$ [96]. High modulus (11 GPa $\leqq$ E $\leqq$ 22 GPa) and strength (T $= 0.34$ GPa) result from these highly oriented chain configurations (see Table 9) [96].

Interestingly, the mechanical behaviour of the fibers is strongly correlated to their properties on the molecular level. Table 9 clearly shows a concomitant increase in director alignment and initial modulus with decreasing spinning temperature. Generally, high viscosities are required to maintain the macroscopic alignment while the fibres are cooling [244–247]. Rheological measurements on LCP 4 indicate a strong increase in melt viscosity with decreasing temperature leading to an optimum spinning temperature of 5 K $\leqq T_s - T_m \leqq$ 15 K [165]. Thus, the highest values for macroscopic alignment ($S_{z'z'} > 0.95$) and initial modulus (E $= 22$ GPa) are found for fibers that were spun just 8 K above the melting point (see Table 9). At present, the mechanical data seem to be the best values reported on LC main chain polymers

a Experiment

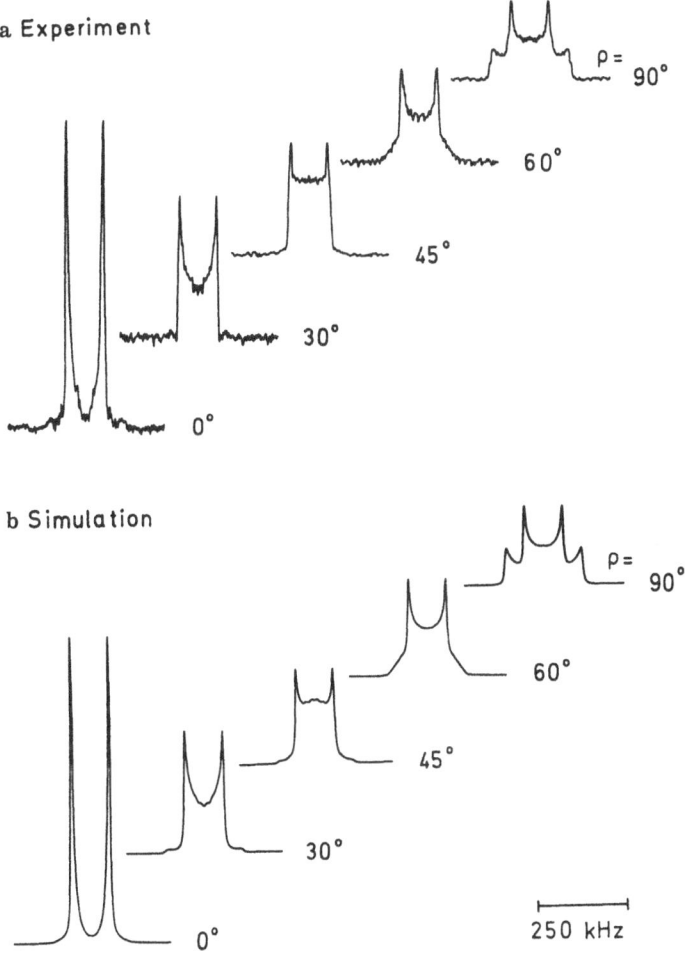

b Simulation

Fig. 30a. Experimental ^{2}H NMR spectra of annealed fibres from main chain polymer **4** (α-CD$_2$) at T = 130 K and five different orientations ϱ of fiber axis and magnetic field. The spectra refer to quadrupole echo sequences (t$_1$/2 = 30 μs). Fiber spining temperature: 458 K. **b)** Simulations were obtained with the parameters given in Fig. 25 and in Tables 3 and 9

containing flexible spacers. Apparently, the high degree of macroorder, maintained in the solid state of these polymers, provides the grounds for their excellent mechanical properties.

Annealing of the LCP fibers only has a minor effect on the mechanical properties [96], in contrast to observations for conventional polymers [17, 248]. This finding can be rationalized by the high degree of order already exhibited by the as-spun fibers [96], which can hardly be improved by any post-processing method [249]. The elongation to break ratios of 2% $\leq \varepsilon \leq$ 3% are considerably smaller than those evaluated for fibers of non-mesomorphic polymers [2]. Apparently, in the LCP fibers the polymer chains are in a highly extended and oriented state (see Table 9), which prevents further drawing.

Table 9. Molecular and mechanical properties of fibers from liquid crystal main chain polymers[a]

Spinning temp.[b] $T_s - T_m/K$	Orders parameters[c]			Mechanical properties[d]		
	$S_{z'z'}$	S_{zz}	$S_{z'z'}$	E/GPa	T/GPa	ε/percent
33	0.80	0.90	0.90	10.7	0.31	2.5
18	0.80	0.90	0.90	11.4	0.32	2.5
8	0.80	0.90	>0.95	22.0	0.34	2.5

[a] LCP **4** with $\overline{M}_n = 30000$.
[b] T_s = spinning teperature, T_m = melting point.
[c] $S_{z'z'}$ = segmental order parameter, S_{zz} = orientational order parameter, $S_{z'z'}$ = macroorder parameter.
[d] E = Young's modulus, T = tensile strength, ε = elongation to break.

Generally, long-range orientational order of LC main chain polymers is quantitatively retained when the sample is cooled below the melting and glass transition, respectively. No change in the order parameters is observed over a period of several years, keeping the sample at room temperature [10, 95, 96]. Therefore, LC main chain polymers can be used as storage materials [250]. Moreover, this particular behavior provides the basis for their numerous applications as high modulus fibers and mouldings [1, 2, 4, 251, 252]. Likewise, main chain polymers have been employed as reinforcing materials in polymer composites and blends [253–257].

In contrast, orientational order of the low molecular weight analogues is completely lost upon crystallization. Thus, many of the exceptional mechanical properties of LC main chain polymers are restricted to systems with higher molecular weights. A prerequisite for any specific applications is precise knowledge of their LC properties at a molecular level. Something, which is now obtainable using dynamic magnetic resonance techniques.

6 Conclusions

Dynamic magnetic resonance, including CW ESR and pulsed ^{2}H NMR, has provided detailed information about molecular order and dynamics of LCPs. The results, referring to side and main chain polymers, have been discussed in relation to the macroscopic properties of these systems. Principal conclusions that emerge from this study are summarized in the following:

(1) Decoupling of main and side chain motions via flexible spacers while effective is not complete. Thus, even in the case of long spacers the viscosities and electro-optical response times of *LC side chain polymers* are at least two orders of magnitude larger than those of the low molecular weight analogues under the same conditions.

(2) The nematic order parameters of *LC side chain polymers*, independent of spacer length and molecular weight, exhibit the same limiting values as the monomeric analogues. Linkage of the mesogenic units to the polymer backbone does apparently not restrict their orientational order, even in the case of a short spacer. However, in contrast to conventional LCs, long-range orientational order is maintained in the solid state.

(3) The macroscopic alignment of *LC side chain polymers* in electric and magnetic fields generally depends on spacer length and molecular weight. Only for sufficient long spacers are the threshold fields and elastic constants comparable to those of conventional LCs. In any case, however, the macroorder achieved can be frozen in at the glass transition. This exceptional molecular property of side chain polymers opens new applications in storage technology and in non-linear optics.

(4) Molecular motions of *LC main chain polymers* in the nematic range are at least two orders of magnitude slower than those observed for the low molecular weight analogues. This is reflected in the bulk viscosities of the polymers, exceeding those of the monomers accordingly. Below the melting point two coexisting phases (LC and crystalline) are observed, differing drastically in their molecular dynamics.

(5) Molecular order of *LC main chain polymers* is characterized by a high degree of conformational and orientational order of the polymer chains. The observed order parameters exceed those of conventional LCs by a considerable amount. The pronounced increase in orientational order from the monomers to the polymers can be rationalized by an intramolecular order transfer via highly extended spacers.

(6) Macroscopic alignment of *LC main chain polymers* is achieved by strong electric, magnetic and shear fields. However, compared to conventional LCs, the required threshold fields are extremely high, implying elastic constants, which exceed those of the corresponding monomers by several orders of magnitude. Generally, the macroorder of the main chain polymers is retained in the solid state, even upon crystallization. This particular behavior provides the basis for their excellent mechanical properties, exploited in various applications as high modulus fibers and mouldings.

In summary, dynamic magnetic resonance in combination with an appropriate relaxation model presents a powerful tool for the molecular characterization of LCPs. Concomitant investigations clearly show that the unique material properties of these polymers are strongly correlated to the specific molecular properties, determined by magnetic resonance techniques. Apparently, the knowledge of the molecular behaviour as a function of the chemical structure provides the scientific basis for a directed design of LCPs with well-defined material properties. In this connection, the improvement of the magnetic resonance techniques is a major challenge in polymer physics. Likewise, the synthesis of new classes of LCPs [29–31, 258–260] with unusual mesophases [261, 262] presents a fascinating problem in polymer chemistry.

Acknowledgements: This review is based on projects of our liquid crystal polymer group at the University of Stuttgart. It is a pleasure to acknowledge colloboration with Professor H. Ringsdorf at the University of Mainz, which is highly appreciated. The authors are grateful to Dr. E. Ohmes, Dr. K. Kohlhammer and A. Schleicher (University of Stuttgart) for their advice and support. Finally we thank Professor R. W. Lenz (University of Massachusetts) for many helpful discussions. Financial support of this work by the Deutsche Forschungsgemeinschaft and Fonds der Chemischen Industrie is gratefully acknowledged.

7 References

1. Dobb MG, McIntyre JE (1984) Adv. Polym. Sci. 60/61: 61
2. Chung TS (1986) Polym. Eng. Sci. 26: 901
3. Koide N (1986) Mol. Cryst. Liq. Cryst. 139: 47
4. White JL, Fellers JF (1978) J. Appl. Polym. Sci. Symp. 33: 137
5. Eich M, Wendorff JH, Reck B, Ringsdorf H (1987) Makromol. Chem. Rapid Commun. 8: 59
6. Monnerie L, Lauprêtre F, Noel C (1988) Liq. Crystals 3: 1013
7. Spiess HW (1983) Coll. and Polym. Sci. 261: 193
8. Berliner LJ (ed) (1976) Spin labeling, Academic, New York
9. Griffin RG (1981) Methods Enzymol. 72: 108
10. Müller K, Meier P, Kothe G (1985) Prog. NMR Spectrosc. 17: 211
11. Gray GW, Windsor PA (eds) (1974) Liquid crystals and plastic crystals, Ellis Horwood, Chichester
12. Kelker H, Hatz R (eds) (1980) Handbook of liquid crystals, Verlag Chemie, Weinheim
13. Demus D, Richter L (1978) Textures of liquid crystals, Verlag Chemie, Weinheim
14. Gray GW, Goodby JW (eds) (1984) Smectic liquid crystals, Leonhard Hill
15. Sackmann H (1984) Progr. Colloid and Polym. Sci. 69: 73
16. Gray GW (ed) (1987) Thermotropic liquid crystals, Wiley, Chichester
17. Ward IM (ed) (1982) Development in oriented polymers, Applied Sci. Publ., London
18. Ciferri A, Krigbaum WR, Meyer RB (eds) (1982) Polymer liquid crystals, Academic, New York
19. Gordon M (ed) (1984) Liquid crystal polymers I–III Springer, Berlin Heidelberg New York (Adv. Polym. Sci. vols 59–61)
20. Chapoy LL (ed) (1985) Recent advances in liquid crystal polymers, Elsevier, Amsterdam
21. Blumstein A (ed) (1985) Polymer liquid crystals, Plenum, New York
22. Finkelmann H, Ringsdorf H, Wendorff JH (1978) Makromol. Chem. 179: 273
23. Ringsdorf H, Schneller A (1981) Br. Polym. J. 13: 43
24. Shibaev VP, Konstromin SG, Platé NA (1982) Br. Polym. J. 18: 651
25. Finkelmann H, Rehage G (1980) Makromol. Chem. Rapid Commun. 1: 31
26. Engel M, Hisgen B, Keller R, Kreuder W, Reck B, Ringsdorf H, Schmidt HW, Tschirner P (1985) Pure and Appl. Chem. 57: 1009
27. Platé NA, Shibaev VP (1987) Comb-shaped polymers and liquid crystals, Plenum, New York
28. Jin JI, Antoun S, Ober C, Lenz RW (1980) Br. Polym. J. 12: 132
29. Kreuder W, Ringsdorf H (1983) Makromol. Chem. Rapid Commun. 4: 807
30. Kreuder W, Ringsdorf H, Tschirner P (1985) Makromol. Chem. Rapid Commun. 6: 367
31. Reck·B, Ringsdorf H (1985) Makromol. Chem. Rapid Commun. 6: 291
32. Shibaev VP, Konstromin SG, Platé NA, Ivanov SA, Vetrov VY, Yakovlev IA (1983) Polym. Commun. 24: 364
33. Simon R, Coles HJ (1984) Mol. Cryst. Liq. Cryst. 102: 43
34. Saupe A (1964) Z. Naturforsch. A 19: 161
35. Wassmer KH, Ohmes E, Portugall M, Ringsdorf H, Kothe G (1985) J. Amer. Chem. Soc. 107: 1511
36. De Gennes PG, (1974) The physics of liquid crystals, Clarendon, Oxford
37. Pincus P (1969) Solid State Commun. 7: 415
38. Emsley JW (ed) (1985) Nuclear magnetic resonance of liquid crystals, D. Reidel, Dordrecht
39. Boyer RF, Keinath SE (eds) (1978) Molecular motions in polymers by ESR, Harwood Acad. Chur
40. McBrierty VJ, Douglass DC (1981) J. Polym. Sci. Macromol. Rev 16: 295
41. Bailey RT, North AM, Pethrick RA (1981) Molecular motion in high polymers, Clarendon, Oxford
42. Spiess HW (1985) Adv. Polym. Sci. 66: 23
43. Mehring M (1983) High resolution NMR spectroscopy in solids, (2nd edn), Springer, Berlin Heidelberg New York
44. Haeberlen U (1976) High resolution NMR spectroscopy in solids, Adv. Magn. Res. Suppl. 1, Academic, New York
45. Farrar TC, Becker ED (1971) Pulse and FT NMR, Academic, New York

46. Spiess HW (1978) In: Diehl P, Fluck E, Kosfeld R (eds) NMR basic principles and progress, vol 15, Springer, Berlin Heidelberg New York, p 55
47. Vold RR, Vold RL (1983) Israel J. Chem. 23: 315
48. Luz Z, (1983) Israel J. Chem. 23: 305
49. Meier P, Ohmes E, Kothe G, (1986) J. Chem. Phys. 85: 3598
50. Abragam A (1961) The Principles of Nuclear Magnetism, Oxford University Press, London
51. Powles JG, Strange JH (1963) Proc. Phys. Soc. 82: 6
52. Woessner DE, Snowden BS, Meyer GH (1969) J. Chem. Phys. 51: 2968
53. Jeffrey KR (1981) Bull. Magn. Reson. 3: 69
54. Noack F, (1986) Prog. NMR Spectrosc. 18: 171
55. Jeener J, Broekaert P (1967) Phys. Rev. 157: 232
56. Spiess HW (1980) J. Chem. Phys. 72: 6755
57. Davies JH, Jeffrey KR, Bloom M, Valic MF, Higgs TP (1976) Chem. Phys. Lett. 42: 390
58. Meier P, Ohmes E, Kothe G, Blume A, Weidner J, Eibl HJ (1983) J. Phys. Chem. 87: 4904
59. Spiess HW, Sillescu H (1981) J. Magn. Reson. 42: 381
60. Torchia DA, Szabo A (1982) J. Magn. Reson. 49: 107
61. Beshah K, Olejniczak ET, Griffin RG (1987) J. Chem. Phys. 86: 4730
62. Vega AJ, Luz Z (1987) J. Chem. Phys. 86: 1803
63. Müller K, Schleicher A, Kothe G (1987) Mol. Cryst. Liq. Cryst. 153: 117
64. Schleicher A, Müller K, Kothe G Liq. Crystals, (submitted for publication)
65. Ernst RR, Bodenhausen G, Wokaun A, (1987) Principles of Nuclear Magnetic Resonance in one and two dimensions, Clarendon, Oxford
66. Millhauser GL, Freed JH (1984) J. Chem. Phys. 81: 37
67. Müller L, Chan SI (1983) J. Chem. Phys. 78: 4341
68. Schleicher A, Müller K, Kothe G (to be published)
69. Luckhurst GR, Poupko R, Zannoni G (1975) Mol. Phys. 30: 499
70. Meier P, Blume A, Ohmes E, Neugebauer FA, Kothe G (1982) Biochemistry 21: 526
71. Luckhurst GR, Poupko R (1974) Chem. Phys. Lett. 29: 191
72. Meirovitch E, Luz Z (1975) Mol. Phys. 30: 1589
73. Meirovitch E, Freed JH (1980) J. Phys. Chem. 84: 2459; (1984) 84: 3281
74. Monnerie L, Lauprêtre F, Noel C (1988) Liq. Crystals 3: 1
75. Atherton NN, (1973) Electron spin resonance, J. Wiley, New York
76. Ernst RR, Anderson WA (1966) Rev. Sci. Instrum. 37: 93
77. Kevan L, Schwartz RN (ed) (1979) Time domain electron spin resonance, J. Wiley, New York
78. Norris JR, Thurnauer MC, Bowman MK (1980) Adv. Biol. and Med. Physics 17: 365
79. Lin T-S (1984) Chem. Rev. 84: 1
80. Stillman AE, Schwartz RN (1981) J. Phys. Chem. 85: 3031
81. Gorchester J, Freed JH (1988) J. Chem. Phys. 88: 4678
82. Kim SS, Weissman SI (1978) Chem. Phys. Lett. 58: 326
83. Levanon H (1987) Rev. Chem. Intermed. 8: 287
84. Fessmann J, Rösch N, Ohmes E, Kothe G (1988) Chem. Phys. Lett. 152: 491
85. Kubo R (1969) Adv. Chem. Phys. 15: 101
86. Freed JH, Bruno GV, Polnaszek CF (1971) J. Phys. Chem. 75: 3385
87. Norris JR, Weissman SI (1969) J. Phys. Chem. 73: 31.19
88. Kothe G (1977) Mol. Phys. 33: 147
89. Muus LT (1972) In: Muus LT, Atkins PW (eds), Electron spin relaxation in liquids, Plenum, New York, p 1
90. Sillescu H (1971) J. Chem. Phys. 54: 2110
91. Flory PJ (1969) Statistical mechanics of chain molecules, Interscience, New York
92. Straley JP (1974) Phys. Rev. A10: 1881
93. Portugall M, Ringdorf H, Zentel R (1982) Macromol. Chem. 183: 2311
94. Müller K (1985) Ph. D. Thesis, University of Stuttgart
95. Kohlhammer K, Müller K, Kothe G (1989) Liq. Crystals, in press
96. Müller K, Schleicher A, Ohmes E, Ferrarini A, Kothe G (1987) Macromolecules, 20: 2761
97. Gordon RG, Messenger J (1972) In: Muus LT, Atkins PW (eds) Electron spin relaxation in liquids; Plenum, New York
98. Moro G, Freed JH (1981) J. Chem. Phys. 74: 3757

99. Kothe G, Wassmer K-H, Ohmes E, Portugall M, Ringsdorf H (1980) In: Helfrich W, Heppke G (eds) Liquid crystals of one- and two-dimensional order, Springer series in Chem. Physics Vol. 11, Springer Berlin Heidelberg New York, p 259
100. Wassmer K-H, Ohmes E, Kothe G, Portugall M, Ringsdorf H (1982) Makromol. Chem. Rapid Commun. 3: 281
101. Meirovitch E, Luz Z (1979) Mol. Phys. 37: 1489
102. Luckhurst GR, Yeates RN (1976) J. Chem. Soc. Faraday Trans. 2 72: 996
103. Finkelmann, H, Benthack H, Rehage G (1983) J. Chim. Phys. 80: 163
104. Finkelmann H (1982) In: Ciferri A, Krigbaum WR, Meyer RB (eds) Polymer liquid crystals, Academic New York, p 35
105. Shibaev VP, Plate NA (1984) Adv. Polym. Sci. 60/61: 173
106. Wasserman AM, Alexandrova TA, Buchachenko AL (1976) Eur. Polym. J. 12: 691
107. Kovarskii AL, Placek J, Szoecs F (1978) Polymer 19: 1137
108. Geib H, Hisgen B, Pschorn U, Ringsdorf H, Spiess HW (1982) J. Am. Chem. Soc. 104: 917
109. Boeffel C, Hisgen B, Pschorn U, Ringsdorf H, Spiess HW (1983) Israel J. Chem. 23: 388
110. Pschorn U, Spiess HW, Hisgen B, Ringsdorf H (1986) Makromol. Chem. 187: 2711
111. Bicerano J, Clark HA (1988) Macromolecules 21: 597
112. Schaefer J, Stejskal EO, Perchak D, Skolnick J, Yaris R (1985) Macromolecules 18: 368
113. Cholli AL, Dumais JJ, Engel AK, Jelinski LW (1984) Macromolecules 17: 2399
114. Kohlhammer K (1989) Ph. D. Thesis, University of Stuttgart
115. Kohlhammer K, Reck B, Ringsdorf H, Kothe G (to be published)
116. Kresse H, Talrose RV (1981) Makromol. Chem., Rapid Commun. 2: 369
117. Kresse H, Kostromin S, Shibaev VP (1982) Makromol. Chem., Rapid Commun. 3: 509
118. Zentel R, Strobel GB, Ringsdorf H (1985) Macromolecules 18: 960
119. Attard GS, Williams G (1986) Liq. Crystals 1: 253; (1986) Polymer Commun. 27: 2
120. Attard GS, Araki K, Williams G (1987) Br. Polym. J. 19: 119
121. Bormuth F-J, Haase W (1987) Mol. Cryst. Liq. Cryst. 148: 1
122. Vallerien SV, Kremer F, Boeffel C (submitted for publ.) Liq. Crystals
123. Casagrande C, Veyssie M, Weill C, Finkelmann H (1983) Mol. Cryst. Liq. Cryst. (Letters) 92: 49
124. Shibaev VP (1988) Mol. Cryst. Liq. Cryst. 155: 189
125. Rupp W, Grossmann HP, Stoll B (1988) Liq. Crystals 3: 583
126. Bock F, Kneppe H, Schneider F (1986) Liq. Crystals 1: 239; (1988) Liq. Crystals 3: 217
127. Coles HJ (1985) Faraday Discuss. Chem. Soc. 79: 201
128. Wendorff JH, Finkelmann H, Ringsdorf H (1978) J. Polym. Sci. Polym. Symp. 63: 245
129. Zugenmaier P, Mügge J (1984) Macromol. Chem. Rapid Commun. 5: 11
130. Zentel R, Strobl GR (1984) Makromol. Chem. 185: 2669
131. Finkelmann H, Day D (1979) Makromol. Chem. 180: 2269
132. Boeffel C, Spiess HW, Hisgen B, Ringsdorf H, Ohm H, Kirste G (1986) Makromol. Chem. Rapid Commun. 7: 777
133. Boeffel C, Spiess HW (1988) In: McArdle CB (ed) Side chain liquid crystal polymers, Glasgow
134. Vasilenko SV, Shibaev VP, Khokhlov AR (1985) Makromol. Chem. 186: 1951
135. Warner M (1988) Mol. Cryst. Liq. Cryst. 155: 433
136. Renz W (1988) Mol. Cryst. Liq. Cryst. 155: 549
137. Noirez L, Cotton JP, Hardouin F, Keller P, Moussa F, Pepy G, Strazielle C (1988) Macromolecules 21: 2891
138. Kirste RG, Ohm HG (1985) Makromol. Chem. Rapid Commun. 6: 179
139. Keller P, Carvalho B, Cotton JP, Lambert M, Moussa F, Pepy G (1985) J. Physique Lett. 46: L1067
140. Moussa F, Cotton JP, Hardouin F, Keller P, Lambert M, Pepy G, Maussac M, Richard H (1987) J. Physique 48: 1079
141. Hardouin F, Noirez L, Keller P, Lambert M, Moussa F, Pepy G (1988) Mol. Cryst. Liq. Cryst. 155: 389
142. Finkelmann H, Naegele D, Ringsdorf H (1979) Makromol. Chem. 180: 803
143. Ringsdorf H, Zentel, R (1982) Makromol. Chem. 183: 1245
144. Talroze VR, Kostromin SG, Shibaev VP, Plate NA, Kresse H, Sauer K, Demus D (1981) Makromol. Chem. Rapid Commun. 2: 305
145. Sefton MS, Coles HJ (1985) Polymer 26: 1319

146. Blinov LM (1983) Electrooptical and magnetooptical properties of liquid crystals, Wiley, New York
147. Saupe A (1960) Z. Naturforsch. A, 15: 815
148. Gruler K, Scheffer TJ, Meier G, (1972) Z. Naturforsch. 27: 966
149. de Jeu WH, Claasen WAP, Spruijt AMJ (1976) Mol. Cryst. Liq. Cryst. 37: 269
150. Coles HJ, Simon R (1984) Mol. Cryst. Liq. Cryst. (Lett.), 102: 75
151. Coles HJ, Simon R (1985) Mol. Cryst. Liq. Cryst. Letters 1: 75
152. Coles HJ, Simon R (1985) Polymer 26: 1801
153. Simon R, Coles HJ, (1986) Liq. Crystals 1: 281
154. McArdle CB, Clark MG, Haws CM, Wiltshire MCK, Parker A, Nesto G, Gray GW, Lacey D, Toyne KJ (1987) Liq. Crystals 2: 573
155. Eich M, Wendorff JH (1987) Macromol. Chem. Rapid Commun. 8: 467
156. Armitage D, Delwart SM (1985) Mol. Cryst. Liq. Cryst. 122: 59
157. Willand CS, Williams DJ (1987) Ber. Bunsenges. Phys. Chem. 91: 1304
158. Eich M, Wendorff JH, Ringsdorf H, Schmidt HW (1985) Macromol. Chem. 186: 2639
159. Zugenmaier P, Mügge J (1985) In: Chapoy LL (ed) Recent advances in liquid crystalline polymers, Elsevier, Amsterdam, p 267
160. Davidson P, Keller P, Levelut AM (1985) J. Physique 46: 939
161. Finkelmann H, Kock H-J, Gleim W, Rehage G (1984) Makromol. Chem. Rapid Commun. 5: 287
162. Schätzle J, Finkelmann H (1987) Mol. Cryst. Liq. Cryst. 142: 85
163. Ferrarini A, Moro G, Nordio PL (1988) Mol. Phys. 63: 225
164. Kohlhammer K (1985) Diploma Thesis, University of Stuttgart
165. Schleicher A (1985) Diploma Thesis, University of Stuttgart
166. Luyten PR, Vold RR, Vold RL (1985) J. Phys. Chem. 89: 545
167. Beckmann PA, Emsley JW, Luckhurst GR, Turner DL, (1983) Mol. Phys. 50: 699
168. Beckmann PA, Emsley JW, Luckhurst GR, Turner DL (1986) Mol. Phys. 59: 67
169. Shimizu H (1962) J. Chem. Phys. 37: 765
170. Ukleja P, Pirs J, Doane JW (1976) Phys. Rev. A14: 414
171. Freed JH (1977) J. Chem. Phys. 66: 4183
172. Dong RY (1983) Israel J. Chem. 23: 370
173. Vilfan M, Kogoj M, Blinc R (1987) J. Chem. Phys. 86: 4183
174. Dippel T, Schweickert KH, Müller K, Kothe G, Noack F (to be published)
175. Müller K, Kothe G (1985) Ber. Bunsenges. Phys. Chem. 89: 1214
176. Wunderlich B, (1980) Macromol. phys. vols 1–3, Academic New York
177. Grebowicz J, Wunderlich B (1983) J. Polym. Sci. Polym. Phys. Ed. 21: 141
178. Stohrer M, Noack F (1977) J. Chem. Phys. 67: 3729
179. Ferry JD, (1980) Viscoelastic properties of polymers, Wiley, New York
180. Boyer RF (1988) J. Polym. Sci. Part B Polym. Phys. 26: 893
181. Müller K, Kothe G (1987) Lect. Notes Phys. 293: 171
182. Schleicher A (1989) PhD. Thesis, University of Stuttgart
183. Butzbach GD, Wendorff JH, Zimmermann HJ (1986) Polymer 27: 1337
184. Wendorff JH, Zimmermann HJ (1986) Angew. Macromol. Chem. 145/146: 231
185. Hentschel D, Sillescu H, Spiess HW (1984) Polymer 25: 1078
186. Dumais JJ, Jelinski LW, Leung L, Gancarz I, Galambos A, Koberstein JT (1985) Macromolecules 18: 116
187. Bovey FA, Jelinski LW (1985) J. Phys. Chem. 89: 571
188. Müller K, Wassmer KH, Lenz RW, Kothe G (1983) J. Polym. Sci., Polym. Lett. Ed. 21: 785
189. Sergot P, Lauprêtre F, Louis C, Virlet J (1981) Polymer 22: 1150
190. Meurisse P, Friedrich C, Dovlaitzky M, Lauprêtre F, Noel C, Monnerie L (1984) Macromolecules 17: 72
191. Lauprêtre F, Noel C, Jenkins WN, Williams G (1985) Farad. Discuss. Chem. Soc. 79: 91
192. Clements J, Humphrey J, Ward IM (1986) J. Polym. Sci., Part B Polym. Phys. 24: 2293
193. Emsley JW, Luckhurst GR, Stockley CP (1982) Proc. R. Soc. London Ser. A 381: 117
194. Samulski ET, (1983) Israel J. Chem. 23: 329
195. Luckhurst GR (1985) In: Chapoy LL (ed) Recent advances in liquid crystalline polymers, Elsevier, Amsterdam p 105

196. Samulski ET (private communication)
197. Sauer TH, Zimmermann HJ, Wendorff JH (1985) Coll. and Polym. Sci. 265: 210
198. Wunderlich B, Grebowicz J (1984) Adv. Polym. Sci. 60/61: 1
199. Hsi S, Zimmermann H, Luz Z (1978) J. Chem. Phys. 69: 4126
200. Abe A (1984) Macromolecules 17: 2280
201. Yoon DY, Bruckner S (1985) Macromolecules 18: 651
202. Auriemmo F, Corradini P, Tuzi A (1987) Macromolecules 20: 293
203. Griffin AC, Britt TR (1981) J. Amer. Chem. Soc. 103: 4957
204. Blumstein RB, Stickless AM (1982) Mol. Cryst. Liq. Cryst. (Lett.) 82: 151
205. Griffin AC, Britt TR (1983) Mol. Cryst. Liq. Cryst. (Lett.) 92: 149
206. Buglione JA, Roviello A, Sirigu A (1984) Mol. Cryst. Liq. Cryst. 106: 169
207. Emsley JW, Luckhurst GR, Shilstone GN, Sage I (1984) Mol. Cryst. Liq. Cryst. (Lett.) 102: 223
208. Ronca G, Yoon DY (1982) J. Chem. Phys. 76: 3295; (1984) J. Chem. Phys. 80: 925
209. Vasilenko SV, Khokhlov AR, Shibaev VP (1984) Macromolecules 17: 2270, 2275
210. Kolinski A, Skolnick J, Yaris R (1986) J. Chem. Phys. 85: 3585
211. Martins AF, Ferreira JB, Volino F, Blumstein A, Blumstein RB (1983) Macromolecules 16: 279
212. Blumstein RB, Stickless EM, Gauthier MM, Blumstein A, Volino F (1984) Macromolecules 17: 177
213. Yoon DY, Bruckner S, Volksen W, Scott JC, Griffin AC (1985) Faraday Discuss. Chem. Soc. 79: 41
214. D'Allest JF, Sixou P, Blumstein A, Blumstein RB, Teixeira J, Noirez L (1987) Mol. Cryst. Liq. Cryst. 155: 581
215. Emsley JW, Luckhurst GR, Shilstone GN (1984) Mol. Phys. 53: 1023
216. Blumstein RB, Poliks MD, Stickless EM, Blumstein A, Volino F (1985) Mol. Cryst. Liq. Cryst. 129: 375
217. Sigaud G, Yoon DY, Griffin AC (1983) Macromolecules 16: 875
218. Griffin AC, Sullivan SL (1988) Polym. Prepr. Am. Chem. Soc. 29: 470
219. Müller K, Kothe G (to be published)
220. Blumstein A, Blumstein RB, Gauthier MM, Thomas O, (1983) J. Asrar, Mol. Cryst. Liq. Cryst. Lett. 92: 87
221. Blumstein A (1985) Polym. J. (Jpn.) 17: 277
222. Griffin AC, Havens SJ (1981) J. Polym. Sci., Polym. Phys. Ed. 19: 951
223. Blumstein A, Thomas O (1982) Macromolecules 15: 1264
224. Roviello A, Sirigu A (1982) Makromol. Chem. 183: 895
225. Müller K, Kothe G (unpublished results)
226. Müller K, Hisgen B, Ringsdorf H, Lenz RW, Kothe G (1984) Mol. Cryst. Liq. Cryst. 113: 167
227. Müller K, Eisenbach C, Schneller A, Ringsdorf H, Kothe G (1984) Prog. Coll. and Polym. Sci. 69: 127
228. Blumstein RB, Stickles EM, Blumstein A (1982) Mol. Cryst. Liq. Cryst. Lett. 82: 205
229. Liebert L, Strzelecki L, van Luyen D, Levelut AM (1981) Eur. Polym. J. 17: 71
230. Mitchell GR, Windle AH (1983) Polymer 24: 1513
231. Noel C, Lauptrêtre F, Friedrich C, Fayolle B, Bosio L (1984) Polymer 25: 808
232. Zheng-Min S, Kleman M (1984) Mol. Cryst. Liq. Cryst. 111: 321
233. Kleman M (1985) Faraday Discuss. Chem. Soc. 79: 215
234. Martins AF, Esnault P, Volino F (1986) Phys. Rev. Lett. 57: 1745
235. Moore JS, Stupp SI (1987) Macromolecules 20: 282
236. Esnault P, Volino F, Martins AF, Kumar S, Blumstein A (1987) Mol. Cryst. Liq. Cryst. 153: 143
237. Blumstein A, Vilasagar S (1981) Mol. Cryst. Liq. Cryst. (Lett.) 72: 1
238. Noel C, Monnerie L, Achard MF, Hardouin F, Sigaud C, Gasparoux H (1981) Polymer 22: 778
239. Maret G, Blumstein A (1982) Mol. Cryst. Liq. Cryst. 88: 295
240. Martin PG, Moore JS, Stupp SI (1986) Macromolecules 19: 2459
241. Blumstein A, Vilasagar S, Ponrathnam S, Clough SB, Blumstein RB (1982) J. Polym. Sci., Polym. Phys. Ed. 20: 877
242. Acierno D, La Mantia FP, Polizotti G, Ciferri A, Krigbaum WR, Kotek R (1983) J. Polym. Sci., Polym. Phys. Ed. 21: 2027
243. Wojtkowski PW (1987) Macromolecules 20: 740

244. Jackson WJ (1980) Br. Polym. J. 12: 154
245. Jackson WJ (1983) Macromolecules, 16: 1027
246. Ide Y, Chung T-S (1984–1985) J. Macromol. Sci. Phys. B23: 497
247. Huynh-Ba G, Cluff EF (1985) In: Blumstein A (ed) Polymeric liquid crystals, Plenum, New York, p 217
248. Ward IM (1975) Structure and properties of oriented polymers, Applied Science Publ., London
249. Stamatoff JB (1984) Mol. Cryst. Liq. Cryst. 110: 75
250. Griffin AC, Hoyle CE, Gross JRD, Venkataram K, Creed D (1988) Makromol. Chem. Rapid Commun 9: 463
251. Jackson WJ, Kuhfuss HF (1976) J. Polym. Sci., Polym. Sci., Polym. Chem. Ed. 14: 2043
252. Economy J (1984) J. Macromol. Sci. Chem. A21: 1705
253. Siegmann A, Dagan A, Kenig S (1985) Polymer 26: 1325
254. Chung T-S, McMahom PE (1986) J. Appl. Polym. Sci. 31: 965
255. James SG, Donald AM, MacDonald WA (1987) Mol. Cryst. Liq. Cryst. 153: 491
256. Bhattacharya SK, Tendolkar A, Misra A (1987) Mol. Cryst. Liq. Cryst. 153: 501
257. Pracella M, Chiellini E, Galli G, Dainelli D (1987) Mol. Cryst. Liq. Cryst. 153: 525
258. Majnusz J, Lenz RW (1985) Eur. Polym. J. 21: 656
259. Ballauff M (1986) Makromol. Chem. Rapid Commun. 7: 407
260. Ringsdorf H, Tschirner P, Hermann-Schönherr, Wendorff JH (1987) Macromol. Chem. 188: 1431
261. Hermann-Schönherr O, Wendorff JH, Ringsdorf H, Tschirner P (1986) Makromol. Chem., Rapid. Commun. 7: 791
262. Ballauff M, Schmidt GF (1987) Mol. Cryst. Liq. Cryst. 147: 163

Plasma Chemistry of Polymers

H. Biederman, Y. Osada
Department of Polymer Physics, Charles University, 18000 Praha 8, Czechoslovakia
Department of Chemistry, Ibaraki University, Mito 310, Japan

This article will describe some of the recent progress in the area of plasma polymerization and plasma treatment. It is not intended to be an exhaustive overview of the field, but instead a summary of the highlights of research studies in this field selected by the authors according to the importance.

Comprehensive reviews on basic phenomena, theory, and reaction mechanisms of plasma-assisted processing and plasma polymerization will be covered in this review. Formation of diamond and amorphous carbon which has attracted considerable attention in the last few years will also be described.

Recent advances in plasma assisted deposition of composite metal/organic (polymeric or carbon) films will be discussed including deposition configurations and processes. Suggested applications particularly in optics and microelectronics will be emphasized.

Possible trends in future research and development of plasma deposition of organic films will also be outlined.

Advances in Polymer Science 95
© Springer-Verlag Berlin Heidelberg 1990

1 Introduction

Plasma chemistry of polymers deals with processes taking place usually in a low temperature plasma generated by glow electrical discharges operated in a molecular gas under low pressure. Such plasma is partially ionized gas consisting of energetic "hot" electrons, "cold" ions, photons, and number of neutrals. The energy transfer from electrons to gas molecules leads to the formation of a host of chemically reactive species such as radicals and ions which along with photons cause a variety of plasma-chemical reactions.

The plasma created by a glow discharge possesses electrons with average energies in the range of 1–30 eV and densitites of 10^9–10^{12} cm^{-3} [1, 2]. The electron temperature T_e, which is related to the electron kinetic energy is not equal to that of the neutral gas T_{gas}. The T_e/T_{gas} ratio is 10^2–10^3 and this type of plasma is called a *non-equilibrium plasma* or *low temperature (cold) plasma*. It is therefore possible for the plasma reaction to proceed at near ambient temperatures (analogous conventional chemical reactions require high temperatures) which is usually very important for organic plasma chemistry.

Chemical reactions in low temperature plasma have been studied extensively over the past 20 years. A number of applications have been proposed in terms of plasma modification of polymers and thin film deposition. Typical uses of plasma treatment and plasma polymerization are listed in Table 1. Several books and review articles covering the subject have been published recently [3, 4]. They contain many references to earlier works [5–9].

Plasma chemistry of polymers may be categorized into four types of reactions:
1. surface reaction of polymers (plasma treatment)
2. plasma polymerization
3. plasma-initiated polymerization
4. plasma reduction.

Table 1. Potential applications of plasma-polymerized films

Electronics	IC, LSI resist, amorphous semiconductor, amorphous fine ceramics etching
Electrics	Insulator, thin film dielectrics, separation membrane for batteries
Chemical processing	Reverse osmosis membrane, permselective membrane, gas separation membran-lubrication, insolubilization
Surface modification	Adhesive improvement, protective coating, abrasion-resistant coating anti-crazing & scratching
Optical	Anti-reflection coating, anti-dimming coating, improvement of transparency, optical fiber, optical waveguide laser & optical window, contact lens
Textile	Anti-flamability, anti-electrostatic treatment, dyeing affinity, hydrophilic improvement, water-repellence shrink-proofing
Biomedicals	Immobilized enzymes, organellas & cells, sustained release of drugs & pesticides, sterilization & Pasteurization, Artificial kidney, blood vessel, Blood bag, Anti-cloting

This article describes some of the recent progress in the area of plasma polymerization and plasma treatment. It is not intended to be an exhaustive overview of the field, but instead a summary of the highlights of research studies in this field as perceived by the authors. Fundamentals of plasma physics and chemistry are also briefly described in order to help understanding of plasma processing in the gas phase and on the substrate.

2 Fundamentals of Plasma Polymerization

2.1 Basic Concepts of Plasma Physics

Plasma polymerization takes place in a low pressure (or low temperature) plasma that is provided by a glow discharge operated in an organic gas or vapor at low pressure. Let us explain the origin of a glow discharge. Consider a diode structure (plane-parallel plate electrode system, e.g., 3 cm in diameter) with a cathode-anode distance of several decimeters, in a glass tube where a gas (e.g., neon) is at a pressure of 133 Pa (1 torr). If we connect this tube in series with a resistance to the dc power supply at a low voltage, we measure very low currents (Fig. 1). Raising the voltage, an abrupt increase of current appears that marks the so-called electric breakdown of the gas.

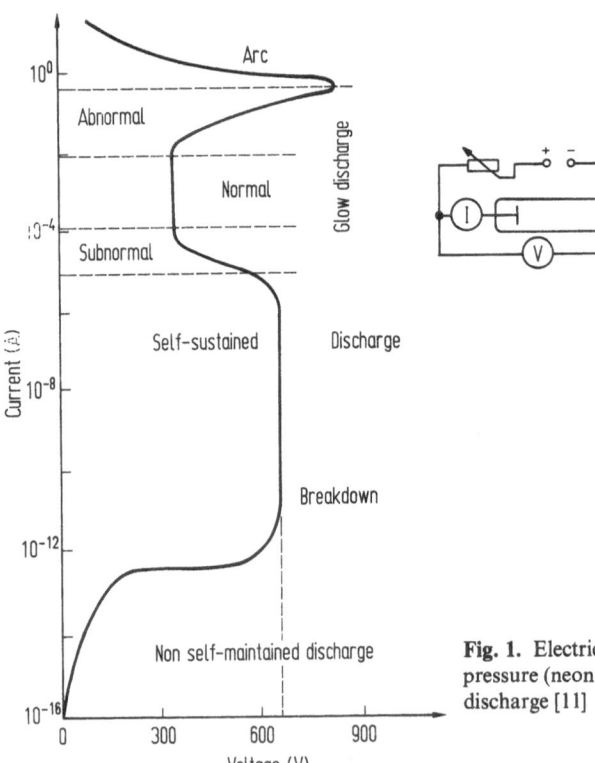

Fig. 1. Electric breakdown of a gas at a low pressure (neon at 1 torr) and formation of a glow discharge [11]

Further increase of the current is only possible when the series resistance is decreased, and we pass subsequently subnormal and normal (discharge current density is constant) glow discharge regions. Further growth of current requires an increase of the applied voltage and we reach the abnormal glow discharge region. At large current values the electrodes are excessively heated. This causes electron thermoemission and the abnormal glow discharge turns into the arc. Generally, normal and abnormal glow discharges are of special importance for thin film deposition techniques including plasma chemistry processes.

The breakdown of a gas in a given apparatus depends on the gas pressure and the inter-electrode separation (corresponding to electric field strength and surface-to-volume plasma ratio). Such relationships are determined experimentally and are known as Pashen's Law (Fig. 2). The physical reasons behind this are the following: the generation of charged particles must be in balance with their loss to walls (these usually predominate) and volume recombination. If the electrode separation is too small, or the gas density (pressure) too low, the electrons are lost without ionizing collisions with the gas atoms. This can be partly compensated by increased voltage. If the pressure (gas density) is too high, the electrons undergo collisions with gas atoms too frequently and cannot accumulate sufficient energy for the ionization. If the inter-electrode distance is too large, then at a given applied potential the local electric field in the plasma is too low to deliver sufficient energy to the electron. In both cases the voltage must be increased. Pashen's law has an important practical meaning because it provides a guidance for the apparatus arrangement, on one hand for stable glow discharge operation and on the other for "shielding distance". Usually the back of electrodes are to be prevented from discharge. This can be done by simply placing closely enough (according to Pashen's Law) a grounded metal shield.

Let us now consider the above-mentioned discharge tube where a normal glow discharge is established in neon at 1 torr pressure (Fig. 3). Historically, dark spaces and luminous regions were observed and named but from the physics point of view four regions are distinguished, e.g. [10, 11]:

I.   cathode region: Aston dark space, cathode layer(s), cathode (Crooke/Hittorf) dark space

II. negative glow region: negative glow, Faraday dark space

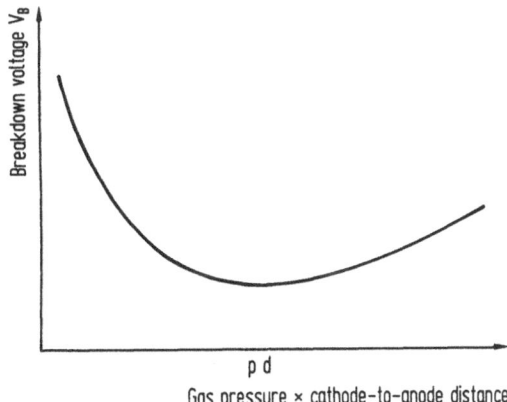

p d
Gas pressure × cathode-to-anode distance

Fig. 2. Pashen's Law (see text) [11]

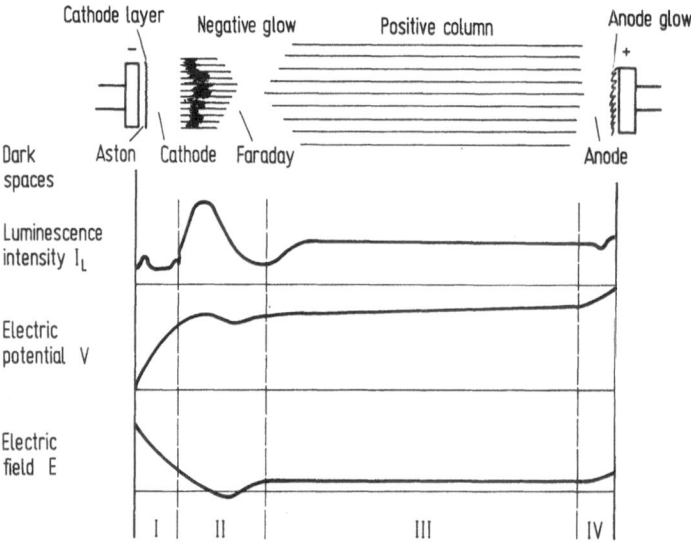

Fig. 3. Normal glow discharge

III. positive column region: positive column

IV. anode region: anode dark space, anode glow layer

The largest potential drop exists over the cathode region which is the most important part for sustaining the discharge. Positive ions are accelerated to the cathode and as a consequence secondary electrons are mitted (JI-process). Electrons are also emitted by energetic photons (JP-process). These electrons are sufficiently accelerated in the cathode dark space so that at its edge inelastic (excitation, ionization) collisions take place. These lead to the creation of positive ions and further electrons. From excited atoms, photons are emitted and therefore from this negative glow the most intensive light irradiation originates. As the electrons pass to the anode they loose energy in inelastic collisions. Ionizing and exciting events subsequently decrease and we approach the Faraday dark space (Fig. 3). Here a large number of low-energy electrons create a negative space charge, so the value of the electric field is very low or even slightly negative. Electrons diffuse along with the ions further towards the anode. This positive column is electrically neutral from a macroscopic point of view (positive and negative particle densities are equal). For this region the term plasma is exactly valid; however, in plasma technology the entire discharge glow is generally understood as a "plasma".

If an ac voltage (up to (kHz) is used, the discharge is still basically of a dc type and each electrode really acts as cathode and anode alternately [12]. Raising the frequency of the applied voltage, positive ions become immobile and the positive space charge is partially retained from one half-cycle to the next (this helps the discharge to re-initiate). An increase of the frequency into the MHz region causes that no significant displacement of either electrons or positive ions happens and losses of charged species from diffusion and recombination processes are replaced by electron-impact ionization of neutral gas molecules in the discharge volume. This type of discharge is called radio frequency (rf) glow discharge (100 kHz — 10 MHz) and it does not depend on processes on the electrodes, which can be covered by any material, e.g., an insulator.

Let us summarize now the basic concepts of plasma physics useful for the description and understanding of plasma chemistry processes.

In a low pressure plasma (glow discharge) the extent of ionization is small, typically 1 charge carrier per 10^6 neutral species (atoms, molecules, etc.).

The negative particles are mostly electrons. However, in the presence of an electronegative gas, e.g., oxygen, a considerable number of negative ions are also formed. In the electric field the electrons gain energy more rapidly than ions. Electrons undergo various collisions with the other particles; that is usually described in terms of a collision cross section (the probability that a given type of collision will occur under given conditions). We may imagine it as a circle of several hundred pm in diameter. The mean free path is a related parameter — it is the average distance traversed by electrons between collisions of a specified type. In the simplest case the mean free path for electrons passing through a gas of particle density n, and producing a certain type reaction, B, during collisions, can be written as:

$$\lambda_B = \frac{1}{n \cdot \sigma_B} \tag{1}$$

where σ_B is the corresponding collision cross section. The total collision cross section can be obtained as the sum of all individual collision cross sections, e.g., elastic, excitation, ionization, attachment, etc.

Let us return to the free electron kinetic energy in a plasma. According to the simple mechanistic model which takes into account only elastic collisions, the free electron kinetic energy should reach very high values even in a very low electric field (E ~ 1 V/cm). Because inelastic collisions take place when the electron energy is >10 eV, the mean electron energy is 3/2 kT ~ 2 — 8 eV. This corresponds to T ~ 10^4 — 10^5 K (simplest case of Maxwellian energy distribution is considered, see Fig. 4). Therefore the electrons are hot. The mean energy for ions is 3/2 kT ~ 0.03 eV which corresponds to T ~ 300 K, approximately the same as for neutral atoms. Therefore these are cold. The presence of hot electrons enables high temperature type reactions (e.g., generation of free radicals) in a low temperature neutral gas.

One of the most important plasma parameters is the collision frequency. For a

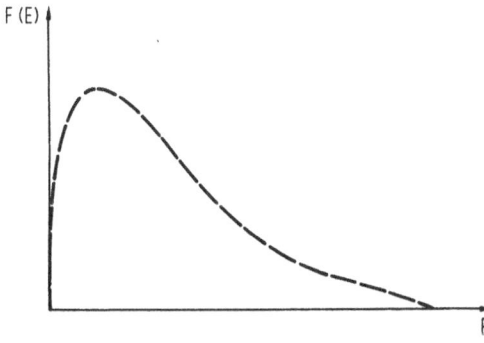

F (E)

E

Fig. 4. Maxwell energy distribution of electrons

particular type of heavy particles (B) the frequency of electron-heavy particle collision (neglecting the velocity of B) can be given by:

$$V_B = n \int_{E=0}^{E=\infty} (E/2me)\ 1/2\sigma_B(E)\ F(E)\ dE \tag{2}$$

where $F(E)$ is the electron distribution function. If this is Maxwellian, then σ_B is energy independent and Eq. (2) reduces to:

$$V_B = n\sigma_B V_e \tag{3}$$

where V_e is the average electron velocity. The expressions for the other types of collision frequencies, as well as for the total collision frequency, are more complicated[13].

A useful quantity, which is directly proportional to the collision frequency, is the reaction rate. For a reaction occurring via a two-body collision between species A and B in a gas at temperature T, the rate can be expressed as:

$$R = k(T)\ N_A N_B \tag{4}$$

where N_A and N_B are the respective particle densities and $k(T)$ is the reaction rate constant. For reactions involving heavy particle collisions the reaction rate constants are used rather than collision frequencies.

The plasma transport properties of various charged species are usually described by drift velocity and drift mobility in the electric field. Correspondingly the electrical conductivity and diffusion coefficients are introduced.

A basic parameter of plasma collective phenomena is the Debye length that provides a measure of the distance over which an extrinsic charge introduced into plasma is shielded [13]. In the simplest case:

$$\Lambda_D \approx = \left(\frac{\varepsilon_0 k T_e}{ne^2}\right)^{1/2} \tag{5}$$

For $n_i \doteq n_e \doteq n \sim 10^{10}\ cm^{-3}$ and $kT_e \simeq 2\ eV$, $\Lambda_D \simeq 105\ nm$. If D is a characteristic plasma reactor dimension, then Λ_D must be $\ll D$, or plasma cannot exist in the case of $D < \Lambda_D$. Electrostatic waves (plasma oscillations) can develop if the electron collision frequency for elastic collisions is considerably lower than the plasma frequency ω_p. Again for the simplest case:

$$\omega_p = \left(\frac{n_0 \cdot e^2}{\varepsilon_0 \cdot m}\right)^{1/2} \tag{6}$$

for $n_0 = 10^{12}\ cm^{-3}$, $f_p \simeq 10\ GHz$, where $f_p = \omega_p/2\pi$ and n_0 is the equilibrium electron density.

The important information for understanding plasma deposition processes is the knowledge of the energy of species bombarding surfaces (reactor walls, electrodes, substrates, etc.) adjacent to the plasma [14]. This may be assessed from the potential variation, which is shown in Fig. 5 for the case of a dc glow discharge deposition sys-

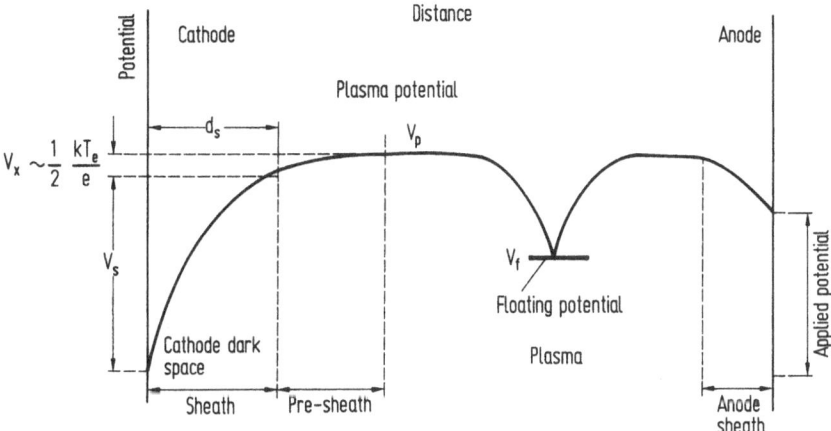

Fig. 5. Distribution of the electric potential in a real dc glow discharge plane-parallel plate electrode system (after Ref. 14, by courtesy of Noyes Publ.)

tem with plane-parallel plate electrodes. The potential of an electrically isolated surface known as the floating potential V_f is given relative to the plasma potential V_p in the simplest case (for electrons with Maxwellian velocity distribution):

$$V_f = \left(\frac{kT_e(eV)}{2e}\right) \ln \left(\frac{\pi m_e}{2M}\right) \tag{7}$$

For Ar, V_f corresponds to about 5 (kT_e) and typical values are about -30 to -40 V. The potential drop between the surface and the plasma is over a certain space, which is called a sheath (typically several Debye length). The nature of the sheath may be explained as follows. Consider a gas of particle density N which performs random motion at temperature T. There is a flux of these particles to the adjacent wall given as:

$$i = \frac{N\bar{v}}{4} \tag{8}$$

where $\bar{v} = (8 kT/\pi m)^{1/2}$ is the average velocity. If we consider now the electrons at temperature $T_e (kT_e \sim 1$ eV) and the positive ions at temperature $T \sim 300$ K, both of typical density 10^9 cm^{-3}, the flux according to Eq. (8) in terms of the electrical current will be of the order of 1 mA/cm^2 and 1 μA/cm^2, respectively. The potential of the surface in contact with the plasma adjusts itself so that the flux of electrons to this surface equals the electron current which is drawn by an external circuit. In the case of an isolated surface its potential reaches the value that equals the electron and ion current to this surface. As a result, the sheath represents a positive space charge. However, there are exceptions, e.g., in the case of a small anode where very high electron current densities are present and a negative space charge sheath is created (Fig. 3). An electric field in the sheath causes coupled diffusion, called ambipolar diffusion, that is described for both electrons and ions by the ambipolar diffusion coefficient.

The largest potential drop occurs in the cathode sheath (cathode dark space) — (Figs. 3 and 5). For the low pressure case, where the ion mean free path is comparable to the cathode sheath thickness, ions will bombard the cathode with an energy corresponding to the cathode potential drop. The ion current density is space-charge limited and is determined by the Child-Langmuir law [14]. At higher pressures (mean free path very small compared to sheath thickness), the charge exchange through ion-atomic collisions becomes important and the bombarding flux will consist of both ions and neutrals with energies considerably less than the cathode potential drop.

In the case of rf discharge (100 kHz — 30 MHz) most of the ions are almost immobile, as we already mentioned, and one would expect only low ion bombardment of the electrodes. However, if one or both of the electrodes are connected to the rf generator via a series capacitor, a pulsating, negative potential will develop on the electrode [15]. The reason for this is the high mobility of the electrons respective to the ions. In the first half of the cycle of the applied rf voltage, a high electron current flows to the electrode and only a relatively small ion current is obtained in the second half-cycle (Fig. 6). As the total ion and electron charge flow to the electrode over an rf cycle must balance to zero, the surface of the electrode self-biases negatively with respect to the plasma potential. The dc component (average dc value) of this potential is nearly equal to the zero-peak voltage applied. The distribution of the potential in an rf glow discharge from the small capacitively coupled electrode (cathode) to the large earthed electrode ("anode"), which is the stainless-steel bell-jar of the reactor, looks the same as in Fig. 5 if we replace $V_s + V_x$ by V_c. Koenig and Maisel [16] have shown that the relation of the voltage between the glow space and the small capaciti-

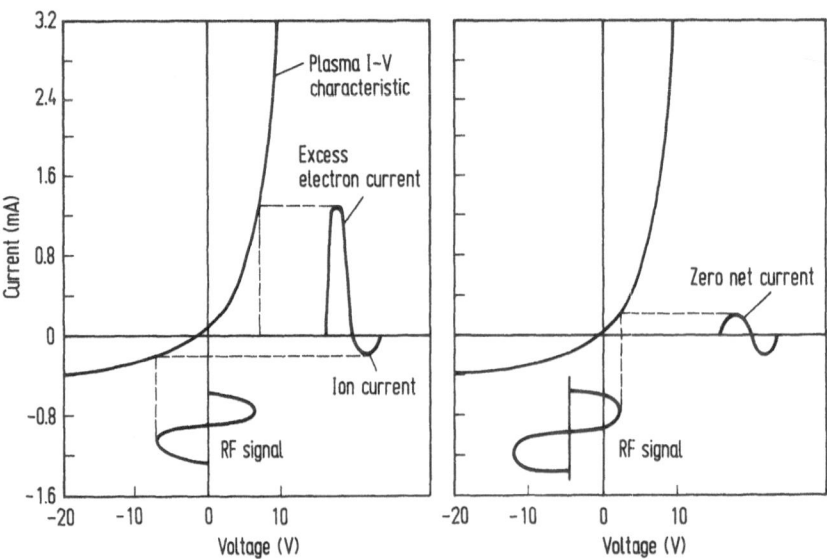

Fig. 6. Negative bias formation on a capacitively coupled surface in an rf glow discharge (after Ref. 15; by courtesy of American Inst. of Physics)

vely coupled electrode V_c to the voltage between the glow space and the large directly coupled electrode V_d is:

$$V_c/V_d = (A_d/A_c)^4 \tag{9}$$

where A_d is the area of the directly and A_c the area of the capacitively coupled electrode. In the case of a tubular reactor made of silica with rf discharge excitation by means of an external coil, the average potential distribution over the chamber diameter is symmetrical. A dc negative self-bias occurs on the inside wall of the reactor (under the coil turn).

2.2 Reactions in a Plasma Volume

An electron with a high kinetic energy, interacting with atoms, has an approximately equal probability of producing either excitation or ionization (excitation is slightly more probable in the case of interaction with a molecule). A semi-classical model of such a collision can be viewed as follows. The electron passing closely by an atom produces in it an electric field due to Coulombic force. This field (its component perpendicular to the electron trajectory) causes a "puls" acting on the atom components. This perturbation of the atom can be theoretically understood as equivalent to the effect of a beam of photons with frequencies corresponding to the Fourier components of the pulse [17].

For selected atoms and molecules Table 2 shows the ionization potentials E_i along with the average energy E_a spent by an electron in creating an electron-ion pair.

The electron interaction with molecules takes place in the same way as described for atoms. The only difference is that the excitation can result in molecular dissociation. Let us give a few examples: First for CF_4, which has the excitation threshold 12.5 eV [18, 19]:

$$e^- + CF_4 \rightarrow [CF_4] + e^- \qquad [CF_4]: \text{excited } CF_4$$

$$[CF_4] \rightarrow CF_3^* + F^*$$

or $\qquad e^- + CF_4 \rightarrow CF_3^* + F^-$

Such reactions are a major source of free radicals and negative ions. These "electron impact dissociations" require less energy than direct ionization events in which only the electrons in the high energy tail of the electron energy distribution take part (cf. Fig. 4). As an example of simple molecular ionization, consider the following reaction:

$$e^- + O_2 \rightarrow O_2^+ + Se^-$$

Also dissociation ionization can happen:

$$e^- + CF_4 \rightarrow CF_3^+ + F^* + 2e^-$$

Table 2. Ionization potential (Ei) and average energy
creating an electron-ion pair

Atom/Molecule	E_i (eV)	E_a (eV)
He	24.58	46
Ne	21.56	37
Ar	15.76	26
H_2	15.43	36
N_2	25.95	36
O_2	12.15	32
CO_2	13.81	34
C_2H_2	11.40	28
CH_4	12.99	29
C_2H_4	10.54	28
C_3H_8	11.15	26
C_6H_6	9.23	27

For the creation of negative ions the attachment of low energy electrons to electronegative molecules is important [20]:

$$e^- + O_2 \rightarrow O_2^-$$
$$O_2^- \rightarrow O^- + O^*$$

Some of these reaction products have long lifetimes (at low pressures), e.g., F atoms
cannot recombine in two-body gas phase collisions into diatomic molecules. Both
energy and momentum conservation principles are not satisfied and a third body is
needed. However, molecular species (e.g., radicals) make a coupling reaction easy
because the dissociation energy can be distributed internally through a large number
of degrees of freedom. An example is:

$$CH_3^* + CH_3^* \rightarrow C_2H_6$$

A last comment will be on metastable species. Atoms and molecules are sometimes
excited into such electronic states from which a radiative transition is not allowed
by quantum mechanical rules. Atoms and molecules in these so-called metastable
states may have very long lifetimes and considerably influence the overall discharge
chemistry. A metastable atom (molecule) can transfer its energy through a collision
with another particle and, if this is of lower ionization potential, the result may be an
ionization or dissociative ionization event. Such processes are known as Penning
ionization processes and may be important under certain conditions in plasma chemistry.

As a summary, the general hierarchy for the production of active species in a molecular gas volume is shown schematically in Fig. 7.

2.3 Polymerization Mechanism

If a suitable organic vapor (monomer) is introduced into a plasma, or when a plasma
of an organic vapor is created, polymerization of the vapor occurs and a polymeric

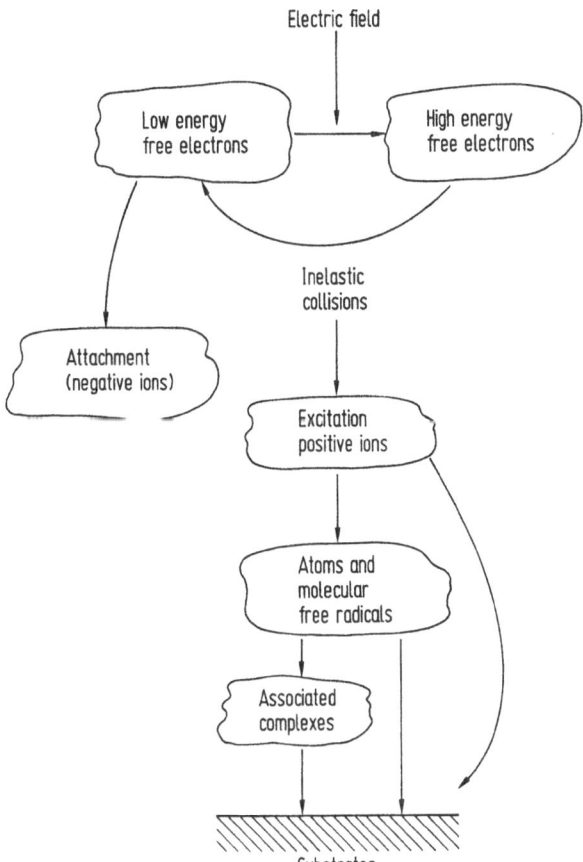

Electric field

Low energy free electrons

High energy free electrons

Inelastic collisions

Attachment (negative ions)

Excitation positive ions

Atoms and molecular free radicals

Associated complexes

Substrates

Fig. 7. The general hierarchy for the production of active species in a molecular gas plasma (after Ref. 14, by courtesy of Noyes Publ.)

film deposits [3, 4, 21, 22]. The method provides a means of surface coating since such a polymer deposition can be highly crosslinked and strongly bonded to the substrate.

Since plasma contains electrons, ions, photons, radicals, and excited molecules, it becomes important to identify the reactive species controlling the propagating process of the polymerization. A number of workers have reported on kinetic models of plasma polymerization. Our current understanding of the chemical and physical mechanism of the process remains limited because the extreme complexity of the plasma environment resists efforts toward a generalization and characterization. The bulk of the research has been concentrated on establishing the dependence of the macroscopic and spectroscopic properties of the product on the major process variables, e.g., rf power, monomer type, and gas flow rate.

Surface reactions are extremely important. They are promoted by bombardment with electrons, ions, and photons. Ions and electrons had been taken into account in the first models of plasma polymerization, e.g., by Williams and Hayes [23], who worked with a 10-kHz frequency applied to parallel plate electrodes. They proposed that the monomer is adsorbed on the surface of the electrodes. By ion and electron bombardment from the discharge, a part of the monomer is converted into surface free radicals that react with the adsorbed monomer molecules causing polymeric chain growth and, finally, creation of a thin hard film.

Later, some other models were proposed based on ion [24] or electron [25] bombardment. Poll et al. [26] also discussed the role of ion bombardment and pointed to a competition between etching and deposition processes in plasma polymerization. A more general description was given by Yasuda [3] (Fig. 8).

Yasuda and co-workers have identified two regimes of plasma polymerization in which the mechanisms differ dramatically: the monomer-deficient and the energy-deficient plasma. In addition, Yasuda has proposed a bicyclic gas phase reaction scheme based in part on his observation of a marked similarity between plasma polymerization and the polymerization of paraxylene. This mechanism postulates a gas-phase of molecules via two cyclic chain growth reaction channels of radicals and di-radicals and the potential attachment of any species to the surface by physical or chemical processes.

Other mechanisms have been proposed which postulate the importance of other species such as ions or neutrals. Based on their observation of deposition rate as a function of saturation of the monomer, Bell and co-workers have suggested that plasma-generated acetylene plays a significant role in film growth when ethane or ethylene was used as a monomer. This mechanism includes the direct attachment of acetylene to the film surface as a major part of the film growth mechanism. Thompson and Mayhan [27] have proposed that ions are the predominant species in film growth, based upon the observation of the effect on film growth of an ion-deflecting element placed in the plasma, and the effect of adding radical scavengers to the plasma.

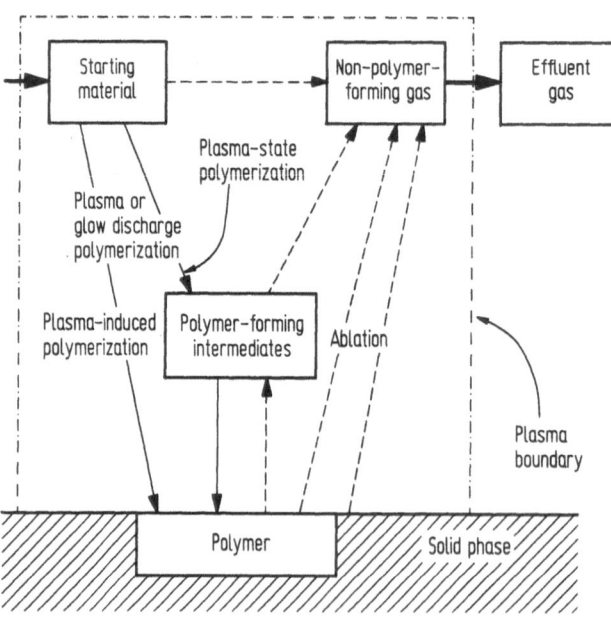

Fig. 8. CAP (Competitive Ablation and Polymerization) (Scheme of glow discharge polymerization by H. Yasuda)

The specific influence of the various reaction parameters, particularly discharge frequency, was studied by S. Morita [28]. He altered the frequency of the plasma in the range from 50 Hz to 13.56 MHz and showed that the rate of polymerization changes in a zig-zag shape and discussed this in connection with the reactive species and their energy which predominate the reaction process. In the case of aromatic hydrocarbons, some authors have tried to differentiate the elementary process of the polymerization between rf and microwave (mw) discharges [29, 30]. M. Duval and A. Theoret [31, 32] investigated the process of polymerization using benzene as monomer. They used the same reactor design for rf and mw and controlled other reaction parameters such as pressure, flow rate, and power. Attention has been given to the formation of oligomers and polymers rather than to the formation of low molecular weight compounds.

Etching processes have been extensively studied by d'Agostino et al. in freon plasmas [33, 34]. It has been well known from reactive ion etching and was also confirmed, e.g., by Biederman et al. [35] in the case of chlorotrifluoroethylene, that from a certain dc negative bias on, the substrate deposition processes are replaced by etching. However, E. Kay [36] reported that in the case of tetrafluoroethylene a solid deposit is obtained up to a negative bias of -600 V. He also mentioned that a mild (~ 10 eV) positive bombardment generally increases the plasma polymerization rate via production of surface free radicals. However, if the negative bias is increased in case of, e.g., hydrocarbons the structure of the polymer film changes and a "hard polymer is obtained" [37]. Holland [38] has shown that if the negative bias is sufficiently high (several hundreds of volts), very hard, so-called amorphous hydrogenated carbon films are deposited. In spite of their partially amorphous structure, these are sometimes called diamond-like carbon (DLC) because of their properties: high electrical resistivity, hardness higher than sapphire, optical transparency in the infrared, and chemical inertness. According to a simple model of Smith [39], these films are composed of diamond, graphitic and polymeric phases (at a short range order), and voids. Weissmantel [40] has shown that a considerable amount of hydrogen (up to 25%) is trapped in the film and that this is connected with excessive compressive stress. Wagner et al. [41] investigated in detail the deposition process using benzene at a pressure of 3 Pa and a dc negative bias $V_B = -1000$ V. Mass spectroscopy has shown that the dominant species among the positive ions is $C_6H_6^+$. Spatially resolved optical emission spectroscopy has shown that $C_6H_6^+$ ions are in part extracted from the plasma glow and accelerated by a sheath potential (close to V_B); at the impact on the substrate they are broken into fragments. The majority of them serve for a-C:H film growth, but part of them is back-scattered. These fragments, e.g., C_2 and especially CH, do not depend on the kind of hydrocarbon employed as the working gas.

Preparative methods for the formation of real microcrystalline diamond films from hydrocarbon/hydrogen mixtures have been developed recently using various types of plasma. However, the roles of plasmas and the growth process of diamond formation have not been well clarified as yet. Tsuda suggested that CH_3^+ ions and H_2 atoms play predominant roles in the formation of the tetrahedral diamond structure [42] (Fig. 9). In that case it seems important to take into account the relative rates of the transportation of radicals containing the carbon fragment and that of the etching or ablation process of graphitic or amorphous deposits by H atoms [43]. Small diamond particles (1—5 µm) were obtained at a rate of 1—5 µm/h from CH_4—H_2—H_2O

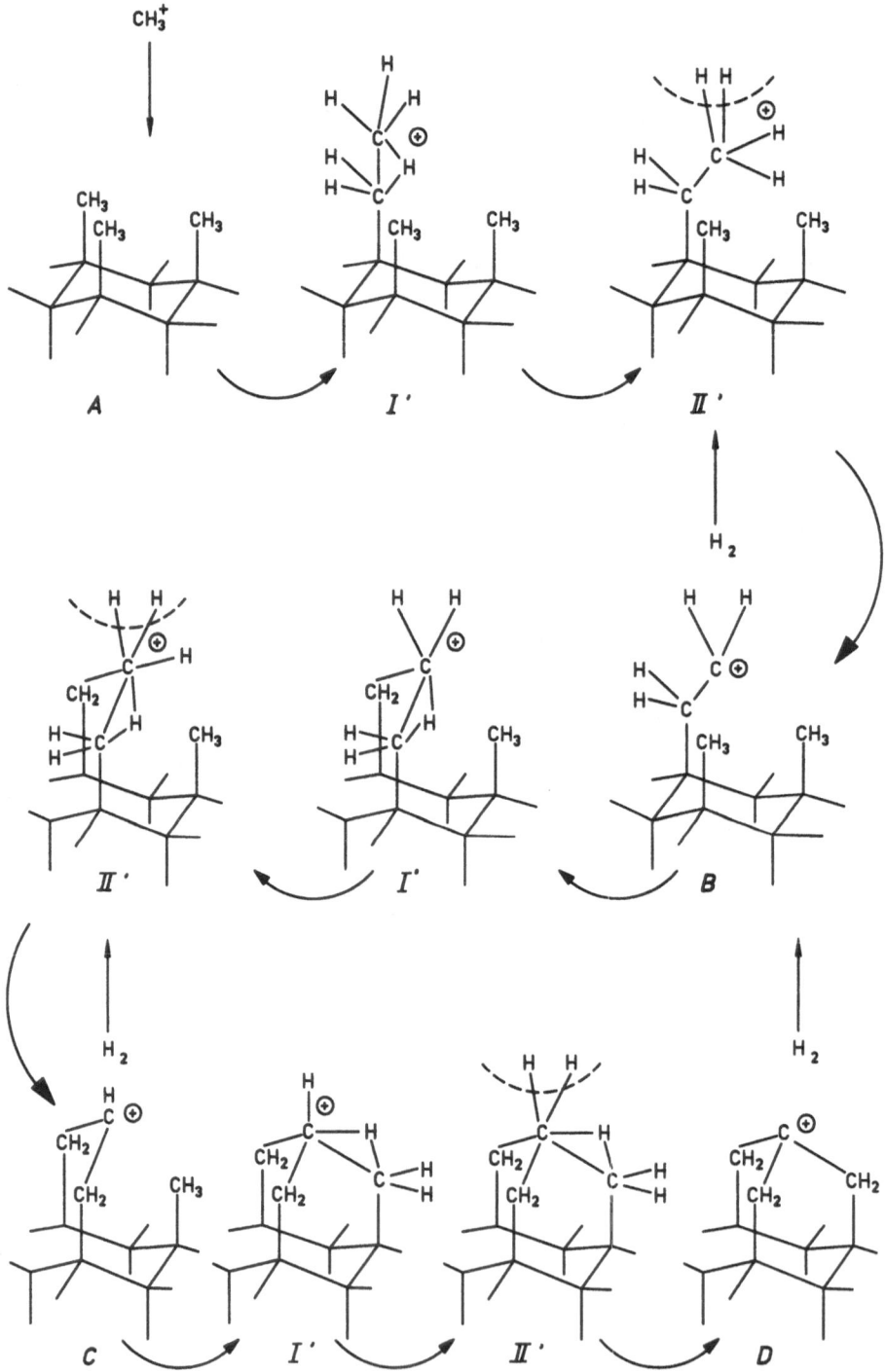

Fig. 9. Postulated mechanism of diamond formation by plasma reaction [42]

mixed gas using microwave plasma. This growth rate was several times faster than that of CH_4-H_2 mixed gas which is the conventional method for diamond preparation by plasma techniques. Promotion of methane decomposition and removal of graphite by-product by OH radicals, generated from H_2O decomposition, were assumed responsible for the increase in growth rate. For a better understanding of the mechanism of the formation of diamond structure, quantitative measurements of the densities and the fluxes of CH_3 and H onto the substrate are required. More details can be found in some review papers, e.g., Refs. [44–46]

Dc glow discharge (post-cathode cylindrical and planar magnetrons) was used for the deposition of a-C:H films [47, 48]. In these cases it is supposed that the cathode is first covered by a-C:H which is sputtered on an electrically floating substrate bombarded simultaneously by energetic neutrals reflected from the cathode. Recently, an "unbalanced planar magnetron" has been employed for hard carbon film production [49, 50].

2.4 Structure of Plasma-Polymerized Films

Plasma-polymerized films are generally highly crosslinked and therefore insoluble in organic solvents. The structure and properties of the products of plasma polymerization vary with changes in plasma parameters; in some cases even soluble oily product and powders are obtained. The films show a complete lack of crystallinity as shown by X-ray diffraction studies. The chemical structure of the polymer is not exactly that of the monomer, largely due to abstraction of hydrogen atoms. Thus, the stoichiometry of the polymers may be strongly changed with the reaction conditions.

The chemical and aggregation structures of plasma-polymerized films are investigated on the basis of thermogravimetric measurements, infrared and ultraviolet absorption spectra, electron microscope, ESR, NMR, and X-ray photoelectron spectra, and others. M. Shen and A. T. Bell quantitatively determined the structure of the plasma-polymerized hydrocarbons, and postulated a molecular model for plasma-polymerized ethylene (Fig. 10). The polymer does not contain regular repeating units of methylene groups, as in conventional polyethylene. Rather it has numerous unsaturated groups, aromatic groups, and side branches. In addition, the polymer is very highly crosslinked, with about one crosslink per six to ten chain carbon atoms. These structural units combine to render the final polymer uncrystallizable [51].

The chemical structure of a plasma-polymerized pyridine film produced on a glass reactor wall by rf power was elucidated by K. Hozumi [52]. As the polymer was highly hydrophilic and was partly soluble in some polar organic solvents, nitrogen-containing polar functional groups were predicted to participate in its chemical structure. ^{1}H-NMR, ^{13}C-NMR, and IR spectroscopy, high-resolution mass spectral data, and number-average molecular weight determination, with some aid from microelemental analysis, revealed in fact the presence of various functional groups such as imine, nitrile, amine, pyridine ring and its N-oxide, and even amide. The authors supposed that the oxygen atom in the last two groups were introduced by contact with ambient air after the plasma process.

The chemical structure of polymerized hexamethyldisiloxane was studied by spectroscopic means such as IR, XPS, and NMR [53]. The plasma polymer was barely

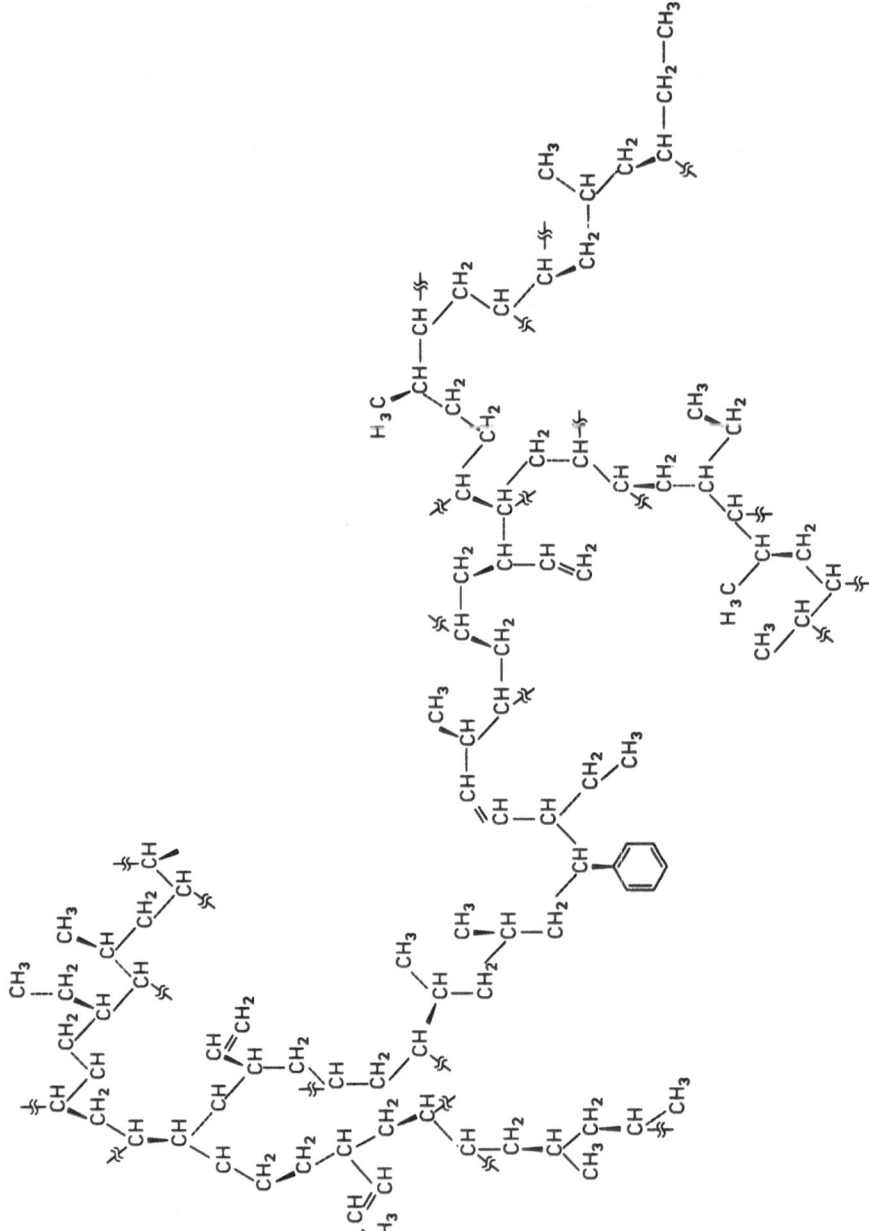

Fig. 10. Postulated model of plasma-polymerized ethylene film [51]

soluble in the usual organic solvents, although it contained a small amount of the monomer and its oligomers. The IR spectrum indicated that the polymer consists of Si—CH$_3$, Si—O, Si—CH$_2$, and Si—H groups. The surface of the polymer retains structural units similar to the monomer as shown by XPS measurements. On the other hand, ^{13}C and ^{29}Si high-resolution, solid-state NMR revealed that the plasma polymer is highly crosslinked with a variety of conformations and a number of O atoms surrounding a Si atom. Results from the XPS and NMR spectra suggested that the bulk of the polymer is more oxidized than the surface layer; the Si atom is preferentially oxidized. A chemical structure was proposed for the polymerized hexamethyldisiloxane.

It is believed that a considerable number of residual radicals are present in the plasma-polymerized thin films, and that these radicals often induce the oxidation of the films to give carbonyls and hydroxyls by contact with the ambient air. The residual radicals formed in plasma-polymerized styrene (PPS) thin films have been investigated by ESR measurements [54]. An asymmetric singlet line characteristic of oxide radicals with anisotropic g-values ($g_1 = 2.008$, $g_2 = 2.006$, and $g_3 = 1.993$) was observed in PPS produced under oxygen carrier gas, whereas a symmetric line was observed in PPS produced under nitrogen carrier gas.

3 Applications of Plasma-Polymerized Organic Films

Research and development of plasma-polymerized organic thin films have been undertaken very actively in the last years and some of the products have been applied in a variety of technologies such as optical, biomedical, electrical, and chemical industries.

3.1 Biomedical Uses

Plasma polymerization is a unique and rather unconventional ultrathin film technology which yields polymers having properties completely different from those of conventional polymers.

Biocompatibility as well as non-thrombogenic properties are closely related to the surface properties of the material. There are several methods to modify the surface properties using physical and chemical reactions, among which the glow discharge polymerization method is one of the easiest ways.

Plasma-polymerized thin film provides a unique opportunity of investigating blood-surface interaction, and according to Alan Hoffman [55], the advantages are: 1) the coating can be applied in such a way that the chemical nature of a surface can be modified without affecting the overall surface topography (thickness of film < 50 nm), 2) the modification of the surface can be achieved without altering the bulk characteristic of the substrate material, 3) the coating is an excellent barrier which prevents low-molecular-weight components of the substrate polymer from leaching, and blood components from penetrating the polymer, and 4) the coating offers imperturbable surface characteristics (in the blood environments). Thus, improvement of the wettability of contact lenses [56] and sterilization of biomaterials [57] by the discharge method have been reported. Yasuda et al. [58], Kronick [59], and Yamashita [60] have reported that plasma polymerization under suitable conditions is effective in improving the thromboresistance of materials. Some results concerning the interaction between platelets and glass surfaces coated with plasma-polymerized thin films from various poly(organosiloxanes) were reported [61]. Organosiloxane polymers were chosen since they are now widely known to exhibit thromboresistance. In this study, the extent of interaction between the prepared materials and blood was evaluated from the number of platelets adhered and the amount of ATP released from the adherent platelets [62], and the effect of plasma-polymerized organosiloxanes as a thrombus-resistant coating was demonstrated.

Tubular blood-contacting polymeric materials were modified by plasma polymerization and evaluated in animals (baboons) with respect to their capacity to induce acute and chronic arterial thrombosis. Nine plasma polymers based on tetrafluoroethylene, hexafluoroethane, hexafluoroethane/H_2, and methane, when deposited on silicone rubber, consumed platelets at rates ranging from $1.1–5.6 \times 10^8$ platelets/cm^2 day. Since these values are close to the lower detection limit for this test system, the plasma polymers were considered relatively nonthrombogenic. Thus, artificial blood tube made of polyesters, having the inner side coated with plasma-polymerized tetrafluoroethylene, is now commercially available.

Plasma polymers of hydrocarbon (CH_4) and perfluorocarbons were inserted into an A–V shunt in a baboon. Clot formation was examined by a r-camera, and the platelet half-life was compared against that of a control experiment according to a previously established standard procedure. Some glow discharge polymers of perfluorocarbons showed excellent athrombogenic characteristics, i.e., no clot formation was observed and platelet half-life was identical to the control value.

Table 3 presents some examples of plasma-treated polymer surfaces prepared for biomedical applications [55]. Additional references on gas discharge-treated biomaterials are cited in the bibliography [1, 3, 4, 55].

3.2 Permselective Membranes

During the past decade a novel technique for the preparation of permselective membranes, including reverse osmosis and ultrafiltration membranes, has been demonstrated; it involves the use of plasma polymerization [1–3, 21, 63]. The composite membranes prepared by the direct deposition of a thin polymer film from the gas phase onto the surface of a porous substrate have a variety of advantages. Plasma polymerization offers a variety of organic thin membranes that reject dissolved ions and organic solutes. Highly salt-rejecting membranes can be formed by the plasma polymerization of a gas mixture of vinylene carbonate and acrylonitrile with deposition of the film on the coarse side of a Millipore filter [64]. However, as with conventional chemically-prepared membranes, a compaction of the membranes was observed with time and the value of water-flux decreased gradually with time. Yasuda [65, 66] concluded from a large number of experiments that an apparent requirement to prepare a semipermeable membrane which allows the transport of water was that the polymer membrane should be moderately hydrophilic. Yasuda [67] and Hollahan [68] indicated that, among organic compounds, nitrogen-containing monomers were particularly suitable for making membranes with high values of salt rejection and water flux.

Table 3. Some examples of plasma discharge-treated polymer surfaces prepared for biomedical applications [55]

Gases (or Monomers)	Polymers	Applications
NH_3 (or $N_2 + H_2$)	Polypropylene (PP) Poly(Vinyl-chloride) (PVC) Polytetrafluoroethylene (PTFE) Polycarbonate (PC) Polyurethane (PU) Poly(methylmethacrylate) (PMMA)	Heparin bonding for improved blood compatibility
Hexamethyldisiloxane (HMDS) $C_2H_4 + N_2$ Allene $+ N_2 + H_2O$	Poly(ethylene terephthalate) (PET) Silastic (SR) Polysulfone (PS)	Improved blood compatibility (in some cases)
Hexamethyl- and Octamethylcyclotetrasiloxane	PP	Improved membrane for blood oxygenator
C_2H_4, allene, styrene, acrylonitrile, C_2F_4, C_2H_3F, C_2F_3Cl, C_2H_3Cl	Polystyrene (PSt) SR	Improved tissue compatibility
C_2H_4, C_2F_3Cl, styrene	SR	Improved tissue compatibility
$C_2H_2 + N_2 + H_2O$	PMMA	Modify corneal contact lens wettability by proteins
C_2H_4, Ar	PP, PET, PVC, SR, Poly(methyl acrylate) (PMA)	Reduce leaching of small molecules from polymer into body
C_2F_4, Et_3SiH, pyridine	PVC (DOP plasticized)	Reduce leaching of DOP into blood
C_2H_4, C_2F_4, C_2H_6, Ar	Poly(2-hydroxyethyl methacrylate) (Poly(HEMA)) Poly(HEMA-MA)	Control of pilocarpine release rate from hydrogel
C_2H_4, C_2F_4, C_2H_6, Ar	SR	Reduce progesterone release rate from SR

The separation of oxygen from air is extremely important in connection with medical treatments, combustion processes, etc. Relatively thin effective membrane layers are required for practical oxygen separation systems, because the permeability coefficients of oxygen of most polymers are lower than those of hydrogen and carbon dioxide.

Plasma polymerization was applied to prepare organic thin films from perfluorobenzene (PFB) [69]. Since the films have a good affinity to oxygen, they were used to separate oxygen gas from air. The aromatic rings changed into highly crosslinked partly saturated structures; the crosslinking increased with increasing magnitude of rf power, substrate temperature, and annealing time, and with decreasing monomer flow rate. The magnitude of the separation factor through plasma-polymerized PFB films deposited on Millipore increased with an increase in the degree of crosslinking as well as thickness (Fig. 11).

Organosilicic compounds are also suitable monomers for highly permeable oxygen separation membranes. Yamamoto et al. used hexamethyldisiloxane as the monomer, yielding composite membranes with both high permeability and high permselectivity [70]. It was demonstrated that the monomer as well as the porous substrate play important roles in the gas permeability of the composite membranes. A compact oxygen enricher was constructed by polymerizing the monomer on a bundle of hollow fibers. Hundreds of the porous glass hollow fibers with a length of about 23 cm were fixed, at one end of each fiber, on an aluminium frame with adhesive tape. The frame was set in a cylindrical chamber and the fibers were suspended in order to distribute each fiber at the same distance from the axis of the cylinder. The efficiency depends on the operating conditions. A pump with a large capacity is necessary to obtain more concentrated oxygen-enriched air. Theoretically, this module can concentrate oxygen to 43%. The average values of P_{O_2} and P_{O_2}/P_{N_2} calculated from the flux and the oxygen concentration of the enricher are shown in Table 4. These values are almost independent of the number of treated hollow fibers as shown in Fig. 12, indicating that the uniformity of the membranes used for the enricher is good.

X-ray photoelectron spectroscopic observations indicate that all plasma polymers contained more than two species of Si atoms, with different oxidation states. The greater part of Si atoms show the same oxidation state as in the corresponding monomer. The gas permeability characteristics are closely related to the oxidation state of the Si atoms in the plasma polymers.

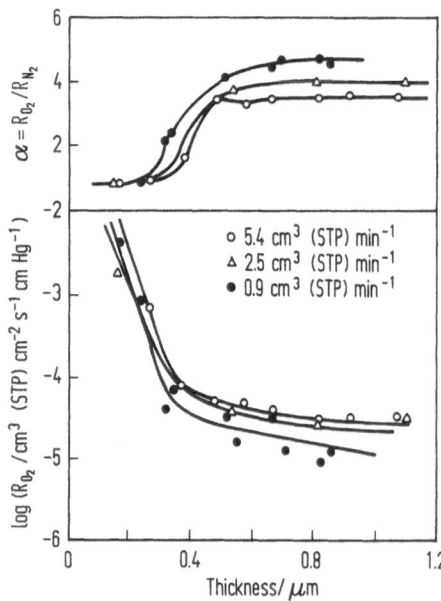

Fig. 11. Thickness dependence of apparent permeatation rate of oxygen, R_{O_2}, and separation factor, α, for plasma-polymerized perfluorobenzene films deposited on Millipore under the conditions of an RF power of 50 W and substrate temperature of 273 K as a function of monomer flow rate [69]

Table 4. Initial efficiency of oxygen enricher[a]

Pressure of permeated gas, $\times 10^4$ Pa	Flux, l/min	Oxygen concentration, %	P_{O_2}[b] $\times 10^{-6}$	P_{O_2}/P_{N_2}
3.07[c]	4.0	34.3	4.5	2.8$_5$
2.13[d]	4.4	36.5		
0[e]	6.8[e]	43[e]		

[a] feed gas is atmospheric air
[b] in cm^3(STP)/cm$^2 \cdot$ sec $\cdot$ Pa
c AULVAC DA-15D vacuum pump was used
d AIWAKI 450D-A vacuum pump was used
[e] theoretical values calculated from the values of P_{O_2} and P_{O_2}/P_{N_2}

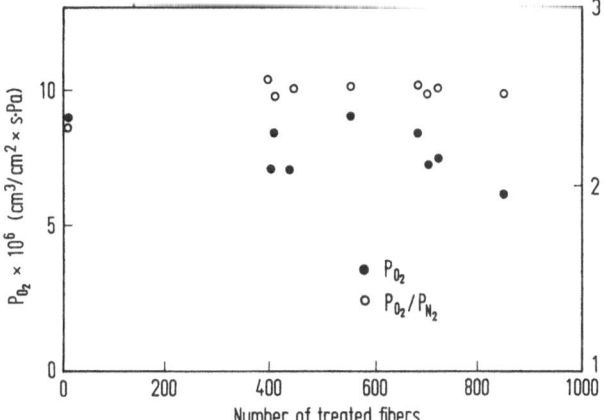

Fig. 12. Influence of the number of treated hollow fibers on the gas permeability of the composite membranes [70]: ($\bullet$) P_{O_2}; (○) P_{O_2}/N_2

The oxygen enricher thus prepared was used for the medical treatment of patients who had lung diseases. It was recognized that the effect of the enricher was equivalent to that of pure oxygen gas at 0.6 l/min. An organosiloxane membrane which can behave as an "elastic valve", changing the pore size reversibly according to the pressure applied, was also prepared [71].

Gas chromatographic research with plasma-coated silica gel adsorbent and an ultrafiltration test of organic solutes having different molecular sizes have revealed that the polymer appears to have micropores of 2–4 nm in diameter. Control of the dissolution rate of pharmaceuticals by changing the polymer film thickness was also attempted [72]. It was found that propargyl alcohol is a very promising monomer and a few applications of this hydrophilic plasma polymer have been reported [72].

Polymeric d-camphor and l-menthol membranes were prepared by plasma polymerization, and permeation and adsorption characteristics of amino acids by these membranes were investigated [73, 74]. Silica particles covered with plasma-polymerized d-camphor membranes on the surface were prepared and adsorption and flow tests of l- and d-tryptophan and some other amino acids were carried out. It was demonstrated that the plasma-polymerized d-camphor membrane permeates d-tryptophan preferably, whereas the plasma-polymerized l-menthol membrane permeates l-tryptophan preferably.

Since dicyclohexyl-18Crown-6 (DC18C6) has the ability to take up ions and to transfer them across a lipophilic medium, it has widely been used in organic synthesis, in ion-extraction into nonpolar solvents, as chiral complexing agents, etc. Osada, Shinkai, et al. have found that membranes prepared under proper conditions retain the structure of the original crown compounds sufficiently to recognize metal ions and v_{K^+}/v_{Li^+} and V_{K^+}/v_{Cs^+} (ratios of ion-permeation rates) for plasma-polymerized membranes were 3.7 and 3.4, respectively [75].

3.3 Protective Coatings

Films fabricated with the plasma polymerization technique have various advantageous characteristics, such as a flawless thin coating, good adhesion to the substrate, mechanical toughness, and thermal stability, and therefore they have a wide variety of potential applications.

T. Wydeven developed a plasma process that significantly improves the hardness of optical lenses and, consequently, leads to greater scratch resistance [76]. Vinyltrimethoxysilane films were deposited on transparent polycarbonate lenses by plasma polymerization. The adherent, clear films protect the substrate from abrasion and also serve as antireflection coatings. Post-treatment of the vinyltrimethoxysilane films in an oxygen glow discharge further improved the abrasion resistance.

A plasma-polymerized tetrafluoroethylene(PPTFE)-coated KBr window transmitted more IR radiation than a commercial antireflective-coated germanium window over the wavelength range 2.5–35 μm [77]. The PPTFE coating also provided some protection against moisture, although the coated windows were unable to meet the humidity specification as given in MIL-C-675A. PPTFE coatings deposited at the downstream edge of the glow provided the best moisture protection.

Exposure of polymeric materials, such as Kapton and Mylar, to the low earth orbit environment of the Space Shuttle has resulted in surface erosion or oxidation. Typical thickness losses for plastic films can be as high as 5–10 μm in a few days. This rapid rate of erosion has generally been ascribed to oxygen atoms impinging on the polymer surfaces at an equivalent translational or kinetic energy of about 5 eV. This observation has prompted studies of the mechanisms of O-atom reactions with polymers and the development of protective coatings [78]. PPTFE was about 7–9 times more resisand to attack by oxygen atoms $O(^3P)$ at ambient temperature than Teflon. Above 393 K, Teflon was more resistant to O-atom erosion than the plasma polymer. A PPTFE coating provided protection for *cis*-polybutadiene against attack by O atoms.

3.4 Electrical Uses

Extensive investigation of the electrical properties of plasma polymeric films using a wide variety of monomers has been carried out and many interesting results have been achieved [79–82]. Also, many studies on the photoconduction of plasma polymeric films have been carried out [83, 84]. Pender and Fleming have obtained plasma polymeric films from styrene, acetylene, benzene, etc., which showed a bistable switching effect [85]. An elastic electron tunneling effect was observed in plasma-polymerized ethylene and benzene films interposed between Al and Pb electrodes at 4 K [86]. In this section only a few examples reported recently are introduced.

Polymer films with electrical conductivities ranging from 10^{-6} to 10^{-7} S cm^{-1} were obtained by the plasma polymerization of fumaronitrile and p-aminobenzonitrile [87]. Al/polymer/Au cells with these polymers showed a rectifying effect. The authors describe that conjugated double bonds formed by plasma polymerization in the heated system were responsible for a relatively high electrical conductivity. The plot of log ι vs. 1/T gave a straight line and an activation energy of 0.29 eV was determined for fumaronitrile polymer.

Tetracyanoquinodimethane (TCNQ) and tetracyanoethylene (TCNE) are known to be strong electron acceptors, and the investigation of their properties is of interest since a number of stable charge transfer complex crystals with high electrical conductivities can be formed from them using a variety of electron donors.

The plasma polymerization of TCNQ and TCNE was carried out at 13.56 MHz from the gas phase and semiconductive polymeric films were obtained [88, 89]. The electrical conductivities of the films obtained ranged from 10^{-10} to 10^{-6} S cm^{-1} and the Al/polymer/ITO (indium tin oxide) sandwich cells made from the films showed rectifying behavior and photovoltaic response. Photoconductivity was also observed in the films. Infrared, ultraviolet, and X-ray photoelectron spectroscopy were utilized to characterize the structure, and these results as well as those from electrical measurements confirmed that a conjugated structure with delocalized π-electrons has been formed in the films.

4 Metal-Containing Organic Thin Films

The incorporation of metals into thin organic polymer films gives rise to a new class of so-called "hybrid films" with interesting physical, chemical, medical, and electrical properties. E. Kay [90–92] introduced the principal methods of metal-incorporation into organic matrices; research of composite plasma polymer films with metal grains dispersed in the organic polymer has since been pursued actively.

4.1 Deposition Techniques

The following methods have been described for the preparation of metal/organic composite films, i.e., metal/plasma polymer or hard carbon.

4.1.1 Simultaneous Plasma Polymerization of an Organic Gas or Vapor and Sputtering or Etching of a Metal from the Target Electrode

This method was first proposed by Kay et al. [90]. An rf diode (parallel-plate electrode arrangement, Fig. 13a) was used. The excitation electrode is coupled via a blocking capacitor, rf matching circuit, rf power meter to one terminal of an rf power supply (usually 13.56 MHz). The other terminal, as well as all metal components of the reactor, are grounded. Plasma polymerization takes place in a glow discharge excited in a monomer gas or vapor, or their mixture with argon, providing thus the dielectric

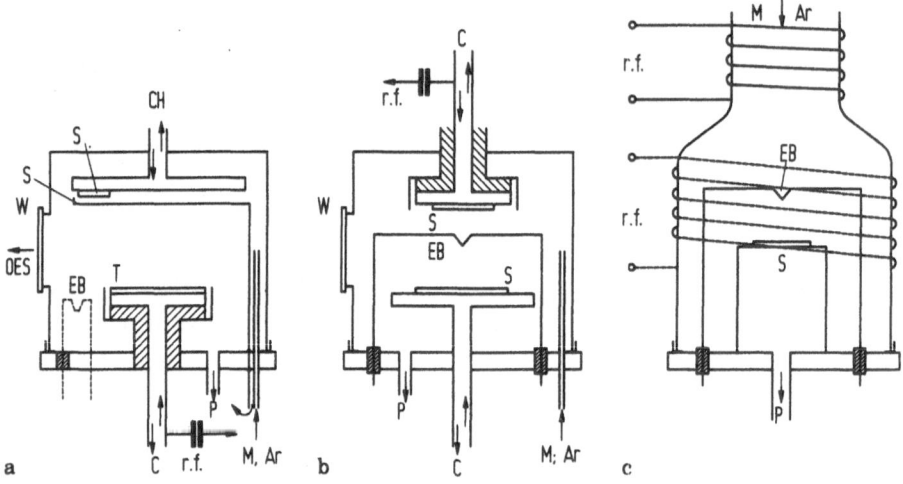

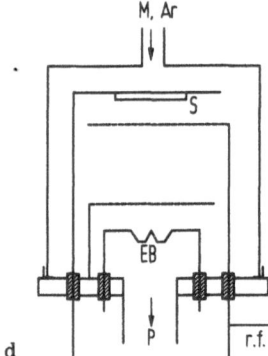

Fig. 13a–d. Deposition arragements: **a.** plane-parallel plate electrode (rf diode) system; plasma polymerization with metal co-sputtering or evaporation. **b.** plane-parallel plate electrode (rf diode) system with a metal co-evaporation for composite plasma polymers (bottom electrode) or a-C:H films (top excitation electrode). **c.** inductive excitation of a monomer or monomer/argon mixture and metal co-evaporation, plasma zone outside-coil: full line; plasma zone around substrate-coil: dashed line. **d.** plane-parallel mesh electrode system with metal co-evaporation.
M — monomer, Ar — argon, S — substrate, Sh — shutter, EB — metal evaporator, rf — to rf power supply, P — to pumps, C — cooling, H — heating, T — metal target

(polymeric) component of the composite. Capacitive coupling of the excitation electrode causes the development of a substantial dc negative self bias V_B (Sect. 2) and therefore sputtering or reactive ion etching of the metal target that supply the metal component of the composite film. Metal emission from the target can be adjusted by adding an appropriate amount of Ar to the monomer gas. However, an increasing concentration of Ar may also influence the polymeric component of the composite. If a planar magnetron is used as the excitation electrode, an increased metal emission and plasma polymerization rate occur at the same power levels.

Several halocarbon gases, such as CF_4, C_3F_8, C_2F_3Cl, C_2F_6, etc. have been successfully applied; for a summary see, e.g., Kay [36] or Biederman et al. [46]. When hydrocarbon (or organosilicon) gases were used, carbonaceous deposits started to cover the excitation electrode and the metal emission from the target was hampered (target poisoning). Janca and Pavelka [93] and Dimigen et al. [94] used even in these cases a dc planar magnetron as the source of metal, but a precise adjustment of the correct ratio of the Ar/organic monomer of the mixture, or a sophisticated spatial distribution of both gases, was necessary.

4.1.2 Simultaneous Plasma Polymerization of an Organic Gas or Vapor and Evaporation of a Metal

An obvious solution to the above-mentioned problems is to use an independent metal supply — a conventional evaporation boat as it is shown in Fig. 13a (Beale [95]). In this deposition system an rf diode is used with the substrate placed on the grounded support. The evaporation boat is positioned close to the target, and sometimes in the middle of the electrodes (Fig. 13b). Using the arrangement shown in Fig. 13a, with a planar magnetron for the discharge excitation, Martinu et al. [96] have shown that the composite films possess identical properties to those prepared in the configuration described in the previous paragraph.

Morita and Hattori [97] used an inductive excitation from outside the reactor chamber made of silica or glass. However, it has been shown by Vossen [98] that even in this case the coupling of the rf power into the discharge is captive and therefore sputter-etching of the inner part of the silica wall, underneath the turn of the coil, can happen and, correspondingly, impurities might be incorporated into the growing deposit (Holland et al. [99]). The substrate may be placed inside or outside the plasma zone as it is shown in Fig. 13c, and the metal is evaporated into the stream of the excited species. Both components travel to the substrate where they combine into a composite film (Hori et al. [100]). In a further development, Itoh et al. [101] used a system where the discharge is excited between mesh electrodes (Fig. 13d). Basically it is a variation of the system from Fig. 13b. A similar system has also been used by Janca and Pavelka [93]. Other workers used instead of mesh electrodes a coil of several turns for the rf discharge excitation (Asano [102] and Suzuki et al. [103]).

The deposition arrangements discussed here may also be used for the deposition of composite films which grow under positive ion bombardment. In the case of the system shown in Fig. 13b, with the substrate placed on the rf powered and capacitively coupled electrode, and using a hydrocarbon gas, one can obtain a hard polymer at a low dc negative self-bias V_B, and a very hard carbon films (a-C:H) at TV$_B$ values higher than -200 V. Using an evaporation boat, the metal can be supplied into the growing film (Biederman et al. [46, 35]). The same procedure can be performed in the case of the system shown in Fig. 13d if a dc negative bias is applied to the substrate holder and conductive substrates are used, or either very thin or highly conducting composite films are deposited. In the case of low conductivity deposits, a better solution is to use an independent rf power supply and capacitive coupling to the substrate holder in order to develop adequate dc negative self-bias. An interesting ion plating system employing a hot filament cathode has been developed by Weissmantel [104] for the deposition of composite hard carbon/metal films. In the case of insulating substrate and low conducting deposits, special precautions must be taken in order to compensate a positive charge build up.

4.1.3 Plasma Polymerization of Metal Organic Compounds

Research in this direction has likely been started in detail by Tkatchuk et al. [105] in the seventies. They reported the structure and properties of organotin polymer films and proposed their application as insulating layers on microelectronic devices, as intermediate adhesive layers, and as protective coatings. Plasma polymerization of tetramethyltin has been further studied by a number of researchers [106–110], with special regard to obtain high conductivity coating usually realized as almost pure tin films. For example, Sadhir and James [106] have reported the synthesis and properties of conducting films with a conductivity as high as 10^4 S cm^{-1}, obtained by means of plasma polymerization of tetramethyltin. The sandwich cells made with Al and Au electrodes showed large and stable rectification effects. Some other studies, e.g., plasma polymerization of ferrocene [111] and iron pentacarbonyl [112], were performed. Recently, an overview of this topic and further results obtained have been published by Oehr and Suhr [113]. They showed that the main process parameters are power density, substrate temperature, and monomer flow. By variation of these parameters it is possible to deposit either metallic films or composite metal/polymer films.

In a more general study Suhr et al. [114] prepared composite metal/polymer films using vapors containing metal organic compounds (of Pd, Ni, Co, Sn, and Au) and various alkenes in a plane-parallel plate electrode reactor (Fig. 13a) with heatable electrodes of 15–25 cm diameter and 2–3 cm apart. Precisely controlled temperature of the substrate resting on the lower grounded electrode is vital, and therefore the authors call the deposition process PECVD (plasma-enhanced chemical vapor deposition). Dc glow discharge, but more conveniently rf (13.56 MHz) plasma excitation, were applied using power levels 0.1–1 W/cm^2. A flow of argon and alkene of 2–100 cm^3 s^{-1} at 20 Pa was used to feed the organometallic compounds from a heatable glass container into the reactor. The organometallic compounds studied were: tetramethyltin, nickeltetracarbonyl, allylcyclopentadienylpalladium, dimethyl-(2,4-pentanedionato)gold (III), and dicobaltooctacarbonyl.

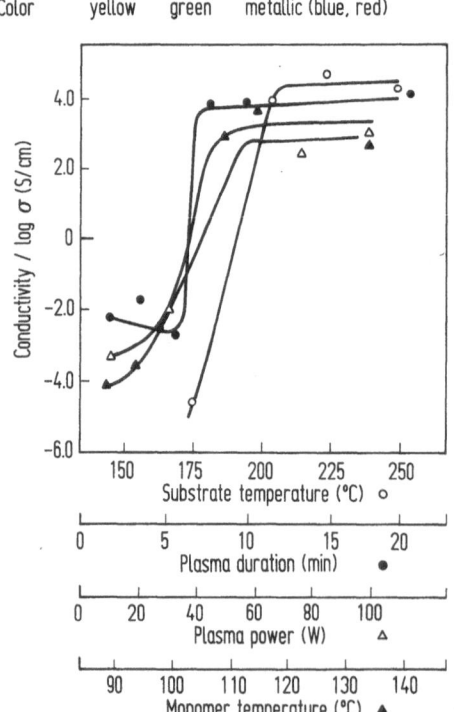

Fig. 14. Relationships of appearance and conductivity with plasma conditions [117, 118]

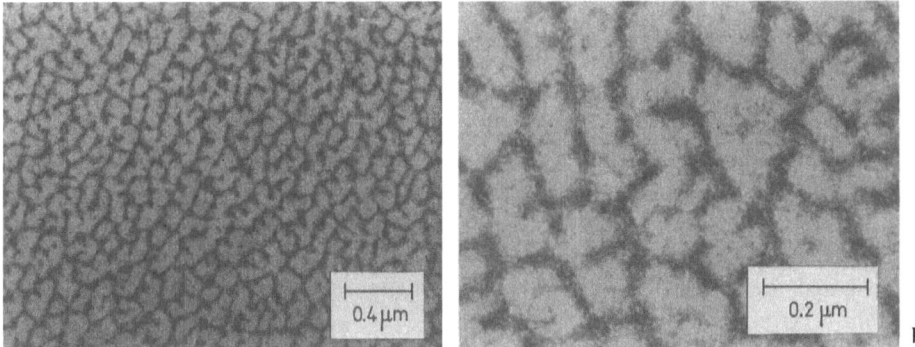

Fig. 15a, b. Transmission electron micrographs of plasma polymerized CuAA film. Conditions of sample preparation [118]: power, 100 W; polymerization, 1.5 min; substrate temperature, 175 °C; magnification, **(a)**, × 30,000, **(b)**, × 1000,000

Polymeric metal-containing (Cu, Mg, Zn, and Ni) and metal-free phthalocyanine thin films were prepared from the gas phase by low-temperature plasma polymerization [115, 116]. The plasma-polymerized CuPc (pp-CuPc) obtained was a glossy greenish thin film with a thickness of 60–300 nm. The film was smooth and even, and was soluble neither in pyridine nor in concentrated sulfuric acid, both of which are good solvents for monomeric CuPc.

Plasma polymerization of copper acetylacetonate (CuAA) was carried out under various conditions and it was found that the color and appearance of the films are determined by the sum of plasma and thermal energies [117, 118] (Fig. 14). The structure and morphology of these films were investigated and it was revealed that the films have a higher-ordered structure consisting of alternating organic and metallic layers (Fig. 15).

Y. Murayama and Y. Yoshida [119] prepared polymeric thin films of dichlorobis (cyclopentadienyl) titanium by the rf ion plating method and investigated the structure and electrical characteristics. The films obtained exhibit a linear increase in electric conductivity with increasing humidity which, according to the authors, has potential application for a humidity sensor.

4.1.4 Sputtering from a Composite Metal/Polymer Target

Composite polymer/metal targets have also been employed following the concept of traditional cermet (cermic-metal) film deposition. Biederman and Holland [120] (using the deposition arrangement pictured in Fig. 13a with a planar magnetron) rf-sputtered, in CF_4, a target made of PTFE with 8 holes in the place of the erosion track where gold was exposed. However, Biederman et al. [121] found that sputtered PTFE in argon is more degraded (deficient in fluorine) than the plasma-polymerized coating. A solution to this problem was suggested by Lehman et al. [122] who recommended to sputter PTFE in a mixture of Ar/CF_4.

Roy et al. [123], using an ordinary rf diode system, sputtered in argon (1 Pa pressure) PTFE/Pt-10% Rh, Au, Cu, and Ekonol/Pt-10% Rh, Au targets. Ekonol (trade name of Carborundum Co) is an aromatic polyester based on *p*-hydroxybenzoic acid.

Varying the target and substrate geometry they succeeded in obtaining various metal contents in the films. They reported cross-contamination effects in the case of PTFE/ metal targets.

Harding and Craig [47], using a dc post-cathode cylindrical magnetron, reactively sputtered stainless steel in an Ar/C_2H_2 mixture. This study was then completed by means of a planar magnetron [124, 48]. Sikkens [125] reactively sputtered composite Ni/C films in a dc diode sputtering system similar to Fig. 13a using an Ar/CH_4 mixture. Weissmantel et al. [126] reported metal/hard carbon films deposited by a dc magnetron co-sputtering of graphite and Ti or Sn. Biederman et al. [50, 127], using an unbalanced planar magnetron with a metal/graphite target, prepared composite films in Ar/propane mixtures. As the metal, mostly silver but also gold, molybdenum, and some others have been examined.

4.1.5 Simultaneous Evaporation of a Metal and a Polymer

Composite films metal/dielectric may be prepared by a simple way: vacuum co-evaporation of both components. "Polymeric films" evaporated in vacuum have been recently reviewed by Hogarth and Iqbal [128]. Boonthanom and White [129], using this approach, deposited composite films using a number of polymers (polyethylene, polycarbonate, melinex, and PVDF) and metals (chromium, aluminium, goled, silver, and copper). Most of the attention was paid to copper/polyethylene films. Solid polymer pellets were evaporated from a special source described by Luff and White [130] at a temperature of 650 K and the metal was evaporated from a molybdenum strip boat. The polymeric component of these films is apparently of lower molecular weight and less crosslinked than it is in the case of plasma polymerization, but nevertheless it is considered of interest to mention this composite film deposition method. The above-cited authors recommend UV light irradiation to improve the properties of the polymeric part of the composite.

4.1.6 Monitoring the Deposition Process

In order to prepare composite films in a reproducible manner, a monitoring technique must be used in situ and it seems that the most simple is emission spectroscopy. A typical spectrum of the near UV and visible regions is given in Fig. 16 for the case of a planar magnetron with gold and PTFE/gold targets sputtered in CF_4 [131]. The most intensive line of gold at $\lambda = 267$ nm may be compared to the height of the band head belonging to the CF_2 radical at $\lambda = 265$ nm. Another example is the deposition of silver/hard carbon films, with monitoring the silver line at $\lambda = 328$ nm in comparison to the argon line at $\lambda = 418$ nm [132]. Quadrupole mass spectrometry may also conveniently be used, but it is more complicated and expensive [36]. Langmuir probes [133] are another simple possibility but one must be cautious to avoid impurities sputter etched from the probes.

4.2 Processes of Deposition and Composite Film Structure

When the deposition techniques described above are applied, the structure of the obtained film is the following: metal grains are embedded in an organic matrix in the

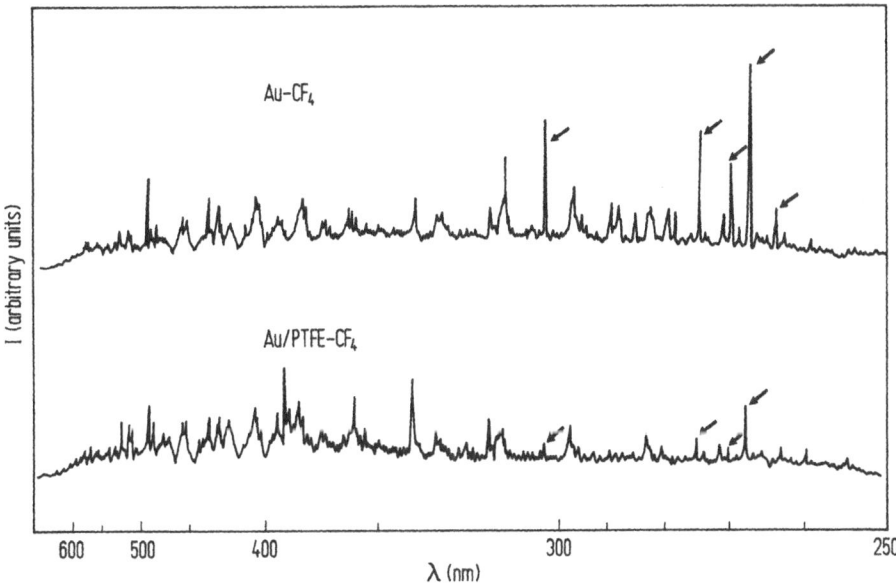

Fig. 16. Spectra of the light emitted from the discharge operated in CF_4 gas when Au and Au/PTFE targets (planar magnetron) were used. Gold peaks are marked by arrows [3]

majority of cases; however, in the case of plasma polymerization of organometallics (or PECVD), the metal may be incorporated on an atomic level. Suhr et al. [114] suggested that the metal precursor is adsorbed at the substrate where it is bombarded by energetic ions, electrons, and photons. If their intensity and duration is sufficient, elemental metal is obtained with all organic parts eliminated. However, if the bombardment is not so intensive and long lasting, a plasma polymer originates from various organometallic and organic moieties. The polymeric part of the composite is strengthened when alkenes are added. The processing parameters, such as the ratio argon/alkene/metal organic in the feed gas mixture and especially the substrate temperature are vital. However, Osada et al. [117] reported an ordinary cluster-like composite film structure for plasma-polymerized CuAA.

Plasma-polymerized halopolymers containing gold attracted considerable attention. These include plasma-polymerized CF_4/Au or CF_4/PTFE-Au systems [131], plasma-polymerized C_3F_8/Au [36, 134–137], and chlorotrifluoro-ethylene(CTFE)/Au [138–140]. The last two composite systems were studied in detail. Figure 17 shows, as an example, electron micrographs of CTFE/Au composite films with increasing gold volume fraction f. These samples were grown on non-thermostated substrates, with increasing rf power to the planar magnetron electrode, with an Au target. The increase of f is accompanied with the enlargement of the gold grain size from 10 to 100 nm at f = 0.5, where among the large grains a system of smaller ones may be seen. An increased grain size was also obtained when the substrate temperature was increased from −10 °C to 60 °C while the other deposition parameters were kept constant [36, 135]. Post-annealing above the temperature of the glass-rubber transition of the plasma polymer (160 °C) caused considerable enlargement of the grain size [36, 136, 137].

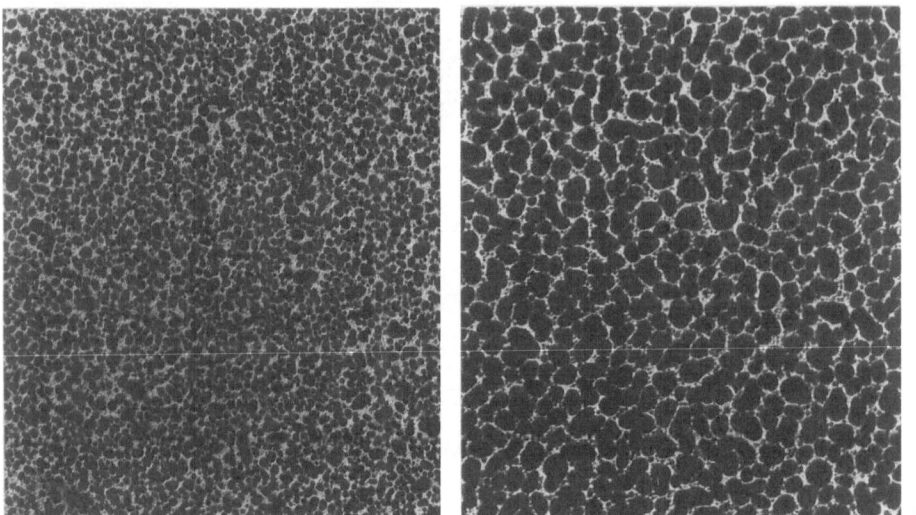

Fig. 17a, b. Structure of fold-doped plasma polymerized CTFE films for samples with **a)** f = 0.11 and thickness 150 nm and **b)** f = 0.37 and thickness 50 nm; observation in TEM [38]

It has already been mentioned that the plasma polymer network is rather disordered and highly crosslinked, with a great number of free radicals incorporated and immobilized. These features may be even more pronounced in the case of composite films where metal atoms are introduced into the growing plasma polymer and create grains (clusters). As a result, one may expect a more disordered structure and a higher concentration of incorporated free radicals. Considerable aging effects in terms of growth of gold clusters have been observed when composite plasma-polymerized CTFE/Au films were exposed to the open atmosphere [141]. The effects involved are likely polymer network rearrangement, when free radicals react with oxygen or water vapor, as well as coalescence caused by diffusion of small gold islands.

AES, ESCA, and IR analyses were employed to examine the chemical composition of gold/halopolymer systems. These techniques show that increased gold content causes a decrease of the halogen concentration, resulting in preferential halogen loss and enhanced chemical reactivity at the surface of the growing layer. In the case of CTFE/Au, AES elemental depth profiling seems to reveal that, as the volume fraction of gold (f) is increased, the surface region of the film is enriched in carbon [46, 139]. The explanation is given in terms of energetic species bombardment (especially negative ions [133]) of the growing composite layer.

Structure and properties of a-C:H films have been mentioned earlier (Sect. 2.3). Weissmantel and co-workers [104, 142] incorporated Cr and Al into an a-C:H matrix using their special "ion plating" system. They found that the increase of metal concentration decreases the hardness of the composite film. Biederman et al. [143], using the rf system shown in Fig. 13c, incorporated Al, Cu, Ag, Ni, and Au by evaporation. Special attention has been devoted to Ag/a-C:H composite films with Ag clusters 3–10 nm in diameter. These films were stable up to ~230 °C and further spherodization took place above this temperature. The hardness of the films again decreased with increasing Ag concentration [132].

4.3 Basic Physical Properties

4.3.1 Optical Properties

Metal/plasma films in most cases (metal clusters embedded in a polymer matrix) behave similarly to cermet films [144, 145]. Anomalous optical absorption in the visible gives rise to their colored appearance. An example may be seen in Fig. 18, and the corresponding structures of the composite films have been shown in Fig. 17. With increasing filling factor f, an anomalous absorption maximum at 550 nm shifts to longer wavelengths and above this point the oberall transmission decreases. In terms of color it is pink for $f \simeq 0.01$, red for $f \simeq 0.06$, violet for $f \simeq 0.24$, and blue for $f > 0.3$. Similar optical properties were reported for plasma-polymerized HMDS/Al, Au, Cu [95, 146], CF_3/Au/Cu [120], and OMTS, HMDS/Au, Cu [93]. However, when reactive metals are chosen as dopants, e.g., Al for halopolymers, the tinting can be caused by the optical absorption of originated chemical compounds [147].

When gold-halocarbon polymer films are annealed, depending on the value of f, the absorption maximum becomes deeper and narrower and transmission at longer wavelengths decreases (for low f) or increases (for high f) [148]. Losses of polymer phase and changes in the gold cluster shape and size (spherodization, coalescence, etc.) are thought to be responsible, especially when the temperature is raised above 160 °C (glass-rubber transition temperature) [136]. Also spontaneous changes were observed after the film was exposed to the open atmosphere, revealing the significance of the aging effects. Coalescence of gold grains was found to be responsible, using direct observation in TEM, as well as theoretical treatment of effective medium theory [141].

Anomalous optical absorption can be theoretically described by the effective medium (EM) approach and investigations in this respect have been carried out in two laboratories; the interested reader can find more details in the literature [136, 140].

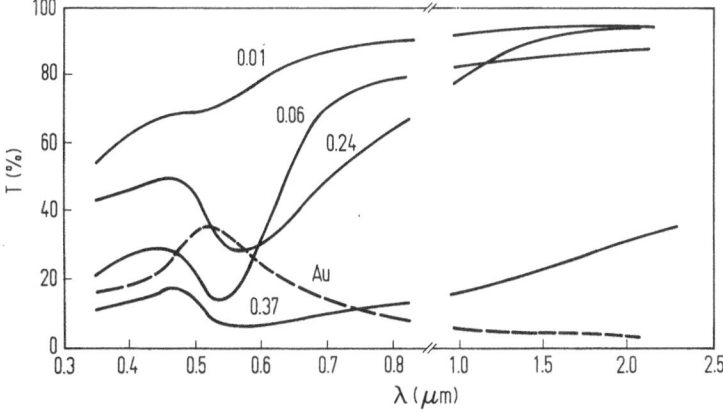

Fig. 18. Spectral dependence of the optical transmission for composite Au/pp CTFE films with the following filling factors and thickness [148]: f = 0.01–198 nm, f = 0.06–183 nm, f = 0.24–29 nm, f = 0.37–76 nm, Au: 28 nm. [148]

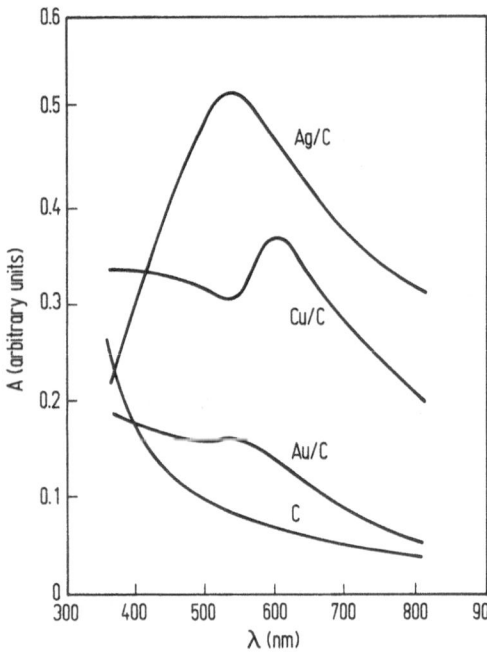

Fig. 19. Absorbance in the visible region of a metal-free a-C : H (C) and a-C : H carbon-doped by silver (Ag/C), gold (Au/C), and copper (Cu/C) [143]

In many cases when the metal/volume fraction (filling factor) f is <0.3, the simple Maxwell-Garnet [149] model described well the reality.

In contrast to plasma polymers, a-C:H films possess a certain intrinsic absorption in the visible region and this gives them a more or less yellowish or brownish color in transmitted light (Fig. 19). This is always superimposed over the anomalous absorption when a metal is incorporated in clusters, so the colors are not so bright. In return one obtains higher thermal stability and better abrasion resistance.

4.3.2 Electrical Properties

Great attention has been paid to the electrical properties in the case of the already mentioned gold/halopolymer films. Electrical resistivity or sheet resistance [96, 120, 131, 137, 150, 151] have been measured as one of the important physical parameters, usually depending on the metal concentration. A percolation threshold (where the conductivity changes several orders of magnitude) was observed at a filling factor f of about 0.4, see e.g. Fig. 20 and a similar curve as that given for the case of pp CuAA films in Fig. 14 results from substituting a power scale for the filling factor scale. Perrin et al.[137], who used a C_3F_8/Ar gas mixture and an rf diode with gold target (Fig. 13a) for their film preparation, found the percolation threshold to be f = 0.37 for as-deposited films and f = 0.42 for films annealed for 2 h at 180 °C. Electric charge carriers (electrons) tunneling between isolated metal grains are supposed to contribute predominantly to the dc conduction. A detailed explanation is given, using the Ping Sheng EM theory, for films above f_p (percolation threshold).

According to the "cermet film" approach, three structural regimes may be distinguished: 1) metallic regime (filling factor is large, metal grains touch and form a metallic continuum with dielectric inclusions); 2) dielectric regime (filling factor is

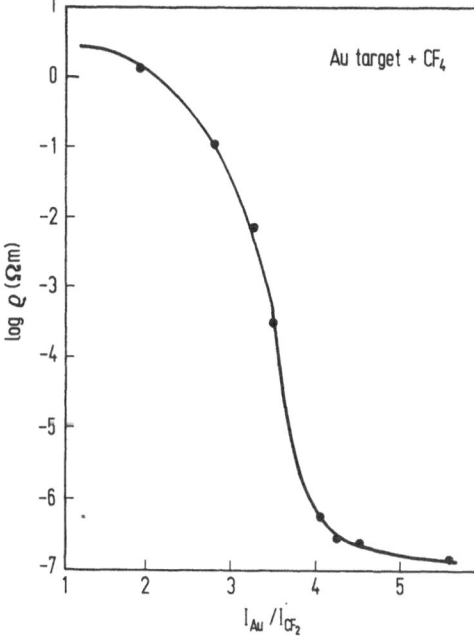

Fig. 20. The resistivity ϱ as a function of the I_{Au}/I_{CF_2} emission intensity ratio which is directly related to the gold content in the film (filling factor f). Abrupt decrease of ϱ corresponds to about f = 0.4 (percolation threshold) [150]

small, isolated metal particles are dispersed in a dielectric continuum); and 3) transition regime (with the percolation threshold f_p).

Suhr et al. [114] reported that, for samples prepared by plasma polymerization of organometallics, no clear percolation threshold was observed in some cases.

The sheet resistance in the case of Au/CF_4 plasma-polymerized films was measured at elevated temperatures. The largest decrease of resistance happens above 150 °C which coincides with the expected glass transition temperature [151]. This irreversible decrease of resistance is likely caused by an overall loss and shrinking of the polymer phase, which allows metal particles to come closer together.

Formerly, Bradley and Hammes [152] studied dc conductivity phenomena on a number of metal-free and some metal-containing plasma polymers. They found plots of lgdc resistivity versus 1/T to be generally linear and concluded that a single activation process is responsible, with activation energies of 1.36 eV for metal-free polymers and <1 eV for metal-doped polymers. For evaporated Cu/polyethylene films [129] the lg sheet resistance revealed a linear dependence on $T^{-1/4}$. The conduction process was explained in terms of hopping of electrons between traps in localized states close to the Fermi level.

Ac conductivity was also shortly studied by some authors [93], in contrast to metal-free polymers where the ac conduction was in the center of attention (for a review see Ref. [153]).

The first report on the electrical properties of "composite hard carbon" films was published by Jones and Stewart [154] who slightly doped their a-C:H films with phosphorus, boron, or nitrogen and found increased conductivity with phosphorus and boron. For the conduction process they proposed a hopping of electrons in a region of high density of localized states for both doped and undoped samples.

Doping or increased substrate temperature during the deposition modified the density of localized states in a way that moved the conduction path closer to the Fermi level fixed with respect to the valence band.

Sikkens [125] found for his composite films (co-sputtered Ni in CH_4/Ar gas mixture) that the resistivity changed little after two hours annealing at 400 °C. Weissmantel et al. [126] measured the dc conductivity of Ti or Sn/hard carbon films as a function of the metal content. Biederman et al. [50] observed aging effects in Ag/hard carbon films prepared by means of the dc unbalanced planar magnetron. In a subsequent study, changes of the electrical resistance after cycling the temperature between liquid nitrogen temperature and 280 °C were observed [127]. Sputtered Ag/graphite in Ar showed a residual increase of resistance (probably Ag coalescence), but Ag sputtered in Ar/propane revelaed a residual decrease of resistance. This is in accord with the idea that the polymeric constituents of this Ag/a-C:H films are partly lost, allowing metal parts to get closer together.

Dc conductivity of Au and Ag/a-C:H films (rf discharge used for the preparation) as a function of temperature, at 10^{-3} Pa, has been studied recently by Biederman et al. [155]. If the maximum temperature did not exceed 260 °C, the lg resistance as a function of 1/T revealed a stable nonlinear form, which is an indication of a non-single activation conduction process. For temperatures lower than 70 °C the lg resistance seems to depend linearly on $T^{-1/2}$ in accord with the theory of Sheng and Abeles [144] for granular metals (also composite metal/dielectric films). When the terminal temperature 370 °C was reached the residual resistance decreased. Again, it seems that polymeric component was lost and the silver percolation maze could therefore improve. Also some of the hard carbon phase might collapse creating graphitic regions or filaments.

A device with a sandwich structure in which the pp-CuPc layer was interposed between Al and ITO electrodes (Al/pp-CuPc/ITO cell) was made. The cell showed rectifying behavior with a forward bias corresponding to a negative voltage on the Al with respect to ITO. It was found that the pp-CuPc exhibits high catalytic activity for photoreduction of solvated methylviologen (MV^{2+}) on irradiation by visible light [116].

Thin films of plasma-polymerized copper acetylacetonate (CuAA) thin films exhibited a wide range of electric conductivities from 10^{-10} to 10^4 S cm^{-1}.

Several workers published "memory switching effects" in metalfree [85, 156–159] and metal-doped polymer or carbon films [117, 158] made in the form of a sandwich metal-polymer-metal. Recently, it has been found that these effects are not likely related to the intrinsic properties of the organic (dielectric) film [157]. More details on composite metal/polymer films can be found elsewhere [159].

4.4 Applications

The first applications proposed for metal/plasma polymer films used their colored appearance. Beale [95] and Wielonski and Beale [146] suggested plasma-polymerized organosilicon films doped with Au, Cu, and Al as decorative coatings. Biederman [131] mentioned the possibility of large optical filters in the case of gold-doped plasma-polymerized fluorocarbon films. Thurstans et al. [161] examined the possibility of using these films as strain gauges. Kay [162] mentioned the use of gold-doped fluoro-carbon films in optical recording. He pointed out that, if the film had a filling factor f lower than the percolation threshold, laser irradiation causes metal coalescence into continuous metal medium with the bulk conduction within spatial dimensions controlled by the laser beam. A pulsed Kr laser at 647 nm with pulse width of 150 nsec caused a sharp optical contrast threshold at 4 mW laser power. Biederman et al. [141] showed that an electron beam may be used with similar results.

An electron beam patterning was also performed on plasmapolymerized CuAA films [117–118]. Morita and Hattori [97] described a dry litography process involving an X-ray mask formed by gold-doped plasma polymerization of styrene.

Composite Te/carbon films were suggested for optical recording by Takeoka et al. [162]. They have shown that these films have higher sensitivity and several times longer life than carbon-free Te films. Electrical and wear abrasive properties of Ta and Sn/carbon films were studied by Weissmantel et al. [126] and suggested for functional electrical coatings in mechanical contacts. Koeberle and Memming [163] incorporated Ta, Ru, W and Ti into a-C:H films with the same intentions. A more detailed study of Ta-doped a-C:H films by Grischke et al. [164] revealed a real possibility to use them for electrical slide contacts. More applications are expected in the future, when a better understanding of the deposition process and the structure of the resulting films will be reached, allowing to prepare metal/polymer and metal/hard carbon composites with reproducible properties and good thermal and time stability.

ZnSe is a semiconductor with a direct band gap of 2.7 eV at room temperature; it has found wide applications for devices such as light emitting diodes (LED$_s$) [165], electroluminescent (EL) cells, [166].

ZnSe thin films were grown by plasma-assisted metalorganic chemical vapor deposition (MOCVD) [167]. As it has already been disscussed in the paragraph 4.1.3 (this process was named PECVD), plasma-assisted MOCUD is a technique combining rf flow discharge decomposition (at 13.56 MHZ) with conventional metalorganic chemical vapor deposition. This process may offer several advantages such as low-temperature growth, high chemical reactivity, cleaning effect of substrate surface, and good surface morphology. The metalorganic sources used were diethylzinc and diethylselenide. Epitaxial films were obtained by this method at a substrate temperature as low as 250 °C. These films were applied in Al/ZnSe:Mn/ITO (indium tin oxide) dc-operated electroluminescent cells. As a manganese source, dicyclopentadienyl manganese (Cp$_2$Mn) was used. This film showed n-type conductivity. The resistivity and carrier concentration were 1.9 Ω cm and 2.4×10^{17} cm^{-3}, respectively. However, the Hall mobility was 14 cm^2/Vs, which is smaller than that of a bulk ZnSe crystal. The authors explained this result by the degradation of crystallinity due to the high-energy ion bombardment. In order to prevent the high-energy ions from damaging the epitaxial layer, the introduction of a mesh electrode between the plane-electrodes as well as control of the distribution of the high-energy ions by applying bias voltage between the mesh and the plane electrode were proposed.

5 Plasma-Initiated Polymerization

5.1 Method and Principle

As described in the previous sections, polymers obtained via plasma are highly branched and highly crosslinked networks and their chemical structure is not simply expressed by that of the starting monomer.

Shen, Bell, and Osada reported a novel polymerization process whereby a low-
pressure cold plasma produced by an electric discharge is used to initiate the polymerization of liquid vinyl monomers [168–170] as well as solid monomers [172–174].
This process is referred to as *plasma-initiated polymerization*, and is characterized by
direct exposure of the liquid monomers to plasma for as short as 60 sec, followed by
post-polymerization at room temperature or below. The resulting polymers are different from common plasma polymers, since they are uncrosslinked soluble polymers,
having extremely high molar masses of 3×10^7 g mol^{-1} or more. The tacticity distribution of the polymer and the monomer-copolymer relationship suggest that the
polymerization is sustained by free radical propagation. However, some questions
arise when the radical mechanism is assumed to account for the polymerization behavior [175]. A.T. Bell et al. [176, 177] have studied the mechanism of plasma-initiated
polymerization in detail; M. Kuzuya [178, 179] postulated a unique process whereby
the growing radicals leached out from the plasma-polymerized films. However,
many points have been left for further study; these are:

1) Why is the polymerization so selective toward the monomer species? Alkyl
esters of methacrylic acid are polymerized by plasma exposure, but styrene (St), α-
methylstyrene (α-MeSt), acrylonitrile (AN), N-vinylpyrrolidone (VPdn), the sodium
salt of p-styrene-sulfonate (NaSS), and other vinyl monomers are not susceptible to
the polymerization.

2) Why is the radical long-lived so as to induce post-polymerization, and how is it
formed? Once the monomer has been exposed to the plasma, it can post-polymerize
for a long period of time. On the base of this nature, it was found that block copolymers are readily prepared using this initiation method [180].

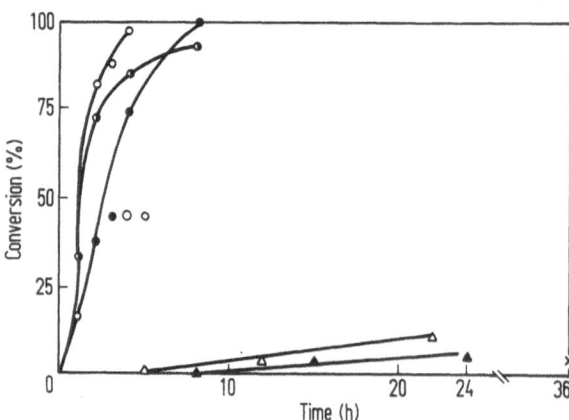

Fig. 21. Time-conversion curves
for the plasma-initiated polymerization of HEMA (◐), AMPS
(●), AAM (○) in water, and
AAM in ethanol (△), DMSO
(▲), and DMF (×); plasma:
100W, 60S; HEMA, 4.15mol/l
(50% v/v); AAM, 4.15 mol/l;
AMPS, 2.08 mol/l; 25 °C post-
polymerization [181, 188]

3) Why is this polymerization process so strongly affected by solvents? What is the role of water in the polymerization?

Figure 21 shows typical time-conversion curves for the plasma-initiated polymerization of acrylamide (AAM), hydroxyethyl methacrylate (HEMA), and 2-acrylamide-2-methyl-1-propanesulfonic acid (AMPS) in water, ethanol, DMSO, and DMF. It is apparent that these monomers polymerize exceedingly fast in water compared with the polymerization in organic solvents. The rate of polymerization of AAM in water was calculated as 4800 times higher than that in DMF. Similar results were obtained for other monomers such as acrylic and methacrylic acids and N-hydroxymethyl-acrylamide [180]. All these monomers were found to polymerize exceedingly fast in water to give extremely high-molecular-weight polymers. For example, poly(AMPS) and poly(AAM) had intrinsic viscosities as high as 15.5 and 22.6 dl g^{-1}, respectively. Poly(AAM) prepared by this method showed a large sedimentation rate against bentonite suspension. Since no such strong solvent effects were obtained by photo-initiated or chemically initiated radical polymerization [182, 183], the results obtained are believed to be inherent to plasma-initiated polymerization. Figure 22 shows the dependence of the rate of polymerization of HEMA on the monomer concentration [175]. The apparent rate of polymerization strongly depends on the water content, with a maximum value obtained at 50% v/v. It was described before that this polymerization may possibly be sustained by free radical propagation. The obtained results, however, seem to be quite unusual for radical polymerization, since in that case a first-order relation-ship with regard to the monomer concentration should hould [184].

Plasma-initiated emulsion (oil in water, water in oil) polymerization of various kinds of vinyl and diene monomers was carried out. Emulsified (0/W) isoprene (I_p) can be polymerized easily by a plasma exposure as short as 60 s, followed by post-polymerization at 25 °C [184]. The obtained polyisoprene had a molar mass of $1.4 \times$

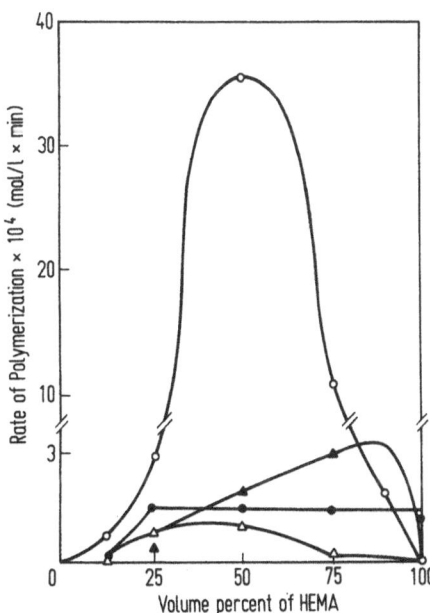

Fig. 22. Monomer concentration dependence in various solvents of the rate of polymerization of 2-hydroxyethyl methacrylate (HEMA) by plasma (○, ●, ☐) and UV (△) exposure [176]. Plasma: 100 W, 60 s; postpolymerization, 25 °C. UV: 400 W mercury lamp; in-source polymerization, 25 °C. Solvent; ○, △, water; ●, ethanol; ☐, DMSO

10^6 g mol^{-1} with 35% cis-1,4-, 55% trans-1,4-, and 10% 3,4-microstructure. The active species generated by this method was long-lived and, therefore, block copolymers of I_p with butyl acrylate and methacrylate could be prepared under emulsion conditions. Water-soluble monomers such as sodium methacrylate, 2-hydroxyethyl methacrylate, and acrylamide were also polymerized rapidly when emulsified in toluene or butyl acetate.

An interesting polymerization process in which plasma is used to generate long-lived active species on the crystals of water-soluble monomers was developed [174]. The active species thus formed on the surface of the crystals cannot initiate the polymerization as long as the monomer remains in the solid state. However, when the monomer crystal is allowed to dissolve by introducing water in vacuo, it starts polymerization spontaneously and very rapidly, to give a polymer with extremely high molecular weight. Thus, once the monomer crystal has been exposed to the plasma for 60 s or less, one can obtain the polymer in situ by merely dissolving the monomer crystal in water (or acid solution in some cases). This polymerization process was referred to as plasma-initiated solvo-polymerization [174].

Crystals of trioxane and tetraoxane can be polymerized in the solid state by plasma initiation followed by post-polymerization. Scanning electron micrographs indicate that PTOX consists of well-aligned fibrils, while PTEOX shows irregular coarse fibers with considerable branching. Small and wide angle X-ray scattering patterns indicate that PTOX crystals obtained through plasma initiation resemble those obtained by γ-ray initiation. However, plasma samples of PTEOX appear to consist of only one rather than two crystalline forms, as shown by both X-ray and differential scanning calorimetric data.

Attempts to polymerize hexachlorocyclotriphosphazene $(PNCl_2)_3$ by high-energy irradiation methods in the solid state have failed to give a significant amount of polymer. Osada, Bell, et al. reported that $(PNCl_2)_3$ crystals can be polymerized quite easily when exposed to a low-pressure radio-frequency plasma [173]. The polymerization takes place almost simultaneously with plasma exposure. The monomer conversion increases with plasma power as well as plasma duration. The product is a white, rubber-like material having an appearance similar to that of the elastomer obtained by thermal polymerization. The IR spectrum shows in fact the same backbone structure. It has also been found that post-polymerization takes place at elevated temperatures following a brief exposure (1–5 min) of phosphazene crystals to plasma. A more detailed kinetic study was carried out recently [187].

5.2 Properties and Applications

The extremely rapid polymerization of monomers in water was applied to immobilize enzymes [186]. Plasma operates essentially at the surface, and a host of chemically active species such as electrons, ions, and radicals in the gas plasma cannot penetrate into the medium more than a few microns [1]. In fact, prolonged plasma exposure scarcely dampened invertase activity [187].

The following are typical characteristics of enzyme immobilization by use of plasma-initiated polymerization:

(i) practically no penetration of energetic plasma into the medium
(ii) brief exposure to plasma
(iii) rapid polymerization at low temperature
(iv) little leakage of enzymes due to ultrahigh-molecular-weight polymers
(v) pure system containing no initiator.

Crosslinked polymer gels prepared by the plasma-initiated copolymerization of AMPS and acrylamide in water exhibited extremely large water adsorption abilities [187]. The amount of adsorbed water and artificial urine attained weights of more than 3000 and 120 times the original weight of dry gel, respectively. Polymer gel obtained by polymerization using radical initiators does not have such high water adsorption abilities. The PAMPS gels prepared by the plasma had also high adsorptive properties versus metal ions such as Cr^{3+} and Co^{2+}. Their binding constants were compared with those of the gels prepared by using radical initiators. Binding constants of PAMPS gels prepared by plasma-initiated polymerization were 35.9 dm^3 mol^{-1} for Cr^{3+} and 30.0 for Co^{2+}, whereas those prepared by conventional radical polymerization were 17.6 for Cr^{3+} and 6.3 for Co^{2+}, respectively [188].

Plasma-initiated graft polymerization of water-soluble vinyl monomers was carried out onto polyethylene, polypropylene, and poly(ethylene terephthalate) films by using the apparatus shown in Fig. 23. It was found that acrylic acid, 2-hydroxyethyl methacrylate, and acrylamide were easily graft-polymerized. Degree of grafting was 0.5–670% of the weight of the dry films, depending on the conditions of plasma-exposure and post-polymerization time. The obtained grafted films showed not only a significant decrease in the contact angle of water, but could adsorb metal ions such as Cu^{2+}, Co^{2+}, and Cr^{3+}. Metal ions were desorbed reversibly by immersing in HCl solution [189] (Table 5).

A composite structure was prepared consisting of a porous substrate membrane on which methacrylic acid (MAA) was graft-polymerized by the method of plasma-initiated polymerization. Using this type of membrane, an effective binding of metal ions is performed by chelation of di- and trivalent metal ions [190]. A pronounced

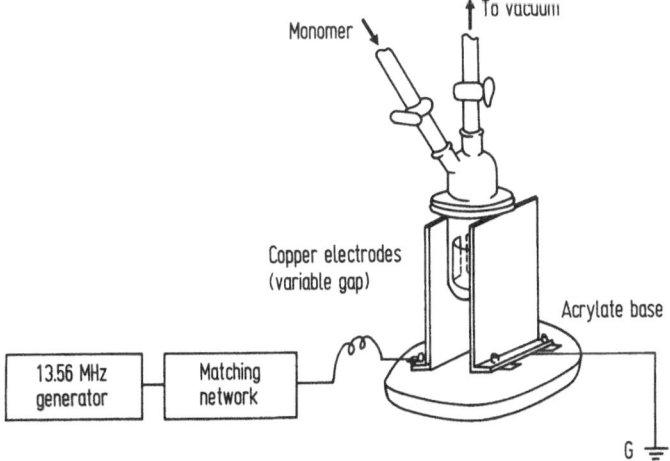

Fig. 23. Apparatus for plasma-initiated graft polymerization

Table 5. Reversible adsorption and desorption of metal ions by plasma-grafted polyethylene films

$\dfrac{[AA]}{[AMPS]}$	Grafting (%)	Adsorbed metal[b] ($\times 10^{-2}$ g metal (g PE)$^{-1}$)					Adsorbed Cu^{2+} [b] ($\times 10^{-4}$ mol g^{-1})	Desorbed Cu^{2+} [c] ($\times 10^{-4}$ mol g^{-1})	Recovery (%)
		Cu^{2+}	Co^{2+}	Cr^{3+}	Cu^{2+}+Cr^{3+}				
3.5	99	3.1	0.7	—	1.5	1.7	2.4	2.3	94
14	360	4.4	2.7	2.8	2.6	2.6	1.5	1.2	76
35	530	5.2	6.7	3.9	4.0	3.9	1.3	1.1	82
7	5	0.6	0.3	—	—	—	1.0	1.1	116
Pure AA	220	1.8	—	0.9	1.3	1.3	0.9	1.1	124

[a] See Table I for conditions of film preparation
[b] Measured by spectrophotometry
[c] Measured by chelate titration

increase in water permeability of the porous membrane was observed due to chelate formation of grafted poly (MAA) with Cu^{2+}.

The increased water permeability can be reversed by removing the metal ions from the membrane. This experiment was performed by alternating addition of Cu^{2+} and ETA (ethylenediaminotetraacetic acid disodium salt, which is an effective polydentate ligand for Cu^{2+}). The result indicates that the water permeability could be varied repeatedly in a range of three orders of magnitude by the alternating addition of Cu^{2+} and ETA at low concentration [190].

5.3 Plasma Reduction

It was found that solid and solvated samples of viologen could be reduced quickly to give the corresponding cation radicals when exposed to a radio-frequency plasma [191, 192].

The plasma exposure was carried out in an optical quartz cell of 10 mm path length, in which 3 cm^3 of a DMF solution of viologen was placed.

The transparent DMF solution of benzylviologen (1,1'-dibenzyl-4,4'-bipyridinium dichloride, BV^{2+}) changed to a deep blue color as soon as it was exposed to the plasma. Figure 24 shows the absorption spectra of plasma-exposed BV^{2+} in DMF at various plasma durations, measured in vacuo. The spectra are in good agreement with the reported spectra of photochemically reduced viologen, showing a monotonous increase of the intensity with the plasma duration. The characteristic absorptions at 402 and 607 nm clearly show the formation of the cation radical BV$^{+\cdot}$. The spectrum disappeared on exposure to air but reappeared with repeated plasma exposure. This fact suggests that the reduction occurred via a one-electron process: BV$^{+\cdot}$ is the major product, since no spectral absorption other than by BV^{2+} and BV$^{+\cdot}$ appear in the UV and visible wavelengths range. Also, no insoluble product was formed. The yield of BV$^{+\cdot}$ based on the amount of BV^{2+} was determined from the ESR spectrum by using a 2,2-diphenyl-1-picrylhydrazyl standard, and was calculated as 66% for a 4.9×10^{-4} mol l^{-1} DMF solution, with plasma exposure for 180 s at 100 W. These facts indicate that the reduction of BV^{2+} is the main pathway.

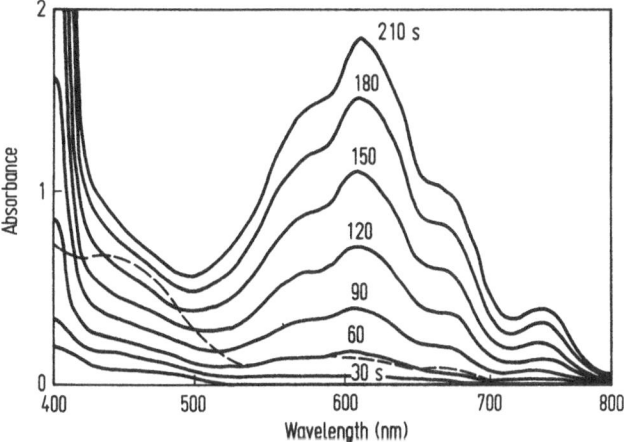

Fig. 24. Progressive spectral changes of benzylviologen with plasma duration [192] (spectra were measured 5 min after plasma exposure). Figures denote plasma duration; after air quenching (———); benzyl-viologen, 4.88×10^{-4} mol/l in DMF (plasma, 100 W)

From experiments at different concentrations of BV^{2+} and different plasma powers, the rate of reduction of BV^{2+} can be expressed as follows:

$$d(BV^{+\cdot})/dt = K(BV^{2+})^{0.5} (power)^{1.0}$$

where K is a constant depending on the plasma reactor.

Methylene Blue (MB) and Brilliant Green (BG), also were subjected to plasma reduction [192]. Figure 25a shows a progressive spectral change of MB with time of

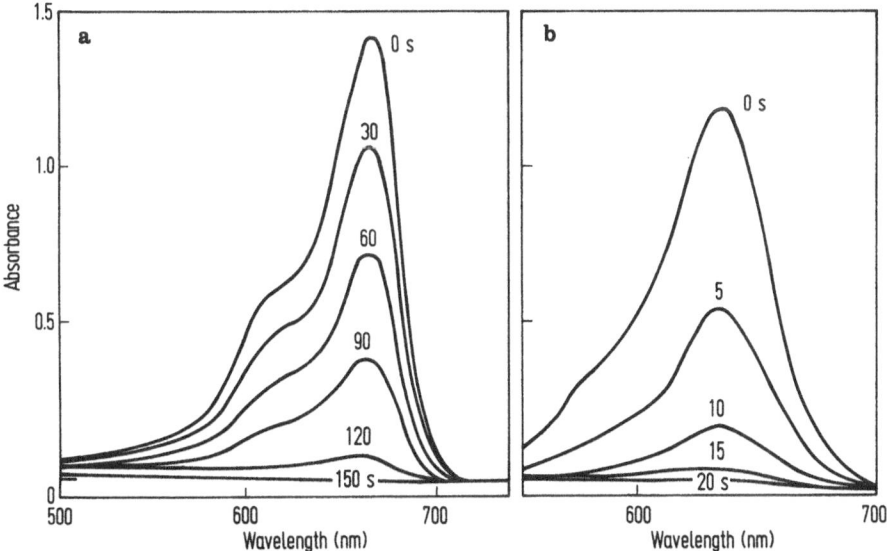

Fig. 25a, b. Progressive spectral changes of MB (a) and BG (b) with plasma duration [193] Figures denote plasma duration (s): MB and BG, 2.0×10^{-5} mol/l in DMF, plasma, 100 W

plasma exposure (plasma duration). Since the absorption of MB at 664 nm was found to decrease in proportion to the plasma duration, the apparent rate of reduction was calculated as 1.62×10^{-7} mol l^{-1} s^{-1}, assuming that the molar extinction coefficient (E) of MB is 71,800. If the process of a plasma reduction of MB is expressed by a simple combination with an electron as:

$$MB^+ + e^- = MB$$

the apparent rate constant of the reduction is calculated as 8.1×10^{-3} s^{-1}. A similar reduction was observed for BG. Figure 25b shows the time profile for the reduction in DMF. The apparent rate constant was calculated as 1.1×10^{-1} s^{-1}, which is about ten-fold larger than that of MB. No insoluble product was formed by the plasma exposure, but in contrast with viologen, only 10–15% of MB reoxidized to the initial form on standing in air for 2–3 days.

Electrons generated by the plasma are quantitatively trapped by DPPH. The

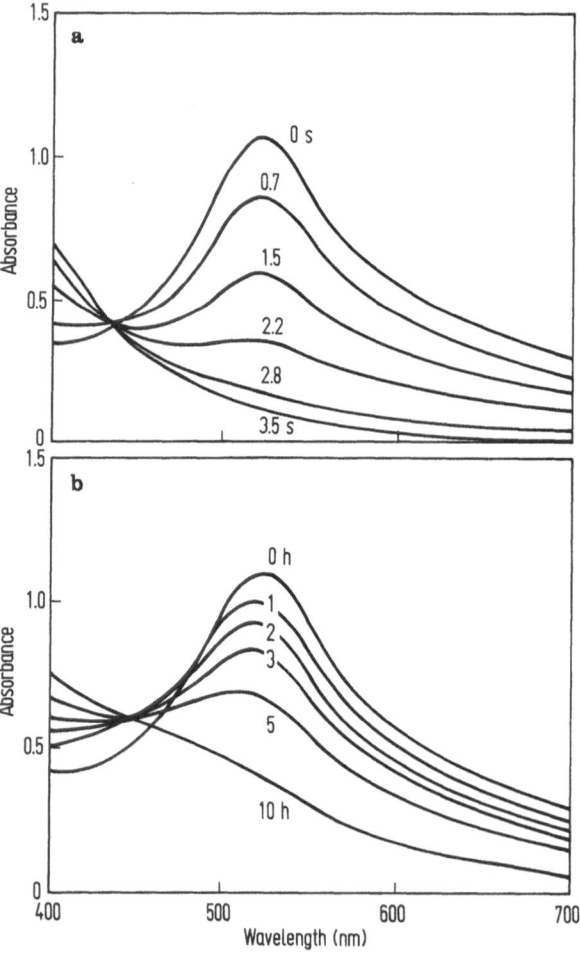

Fig. 26a, b. Progressive spectral changes of DPPH by plasma exposure (a) and thermal decomposition of AIBN (b) in DMF [193]. Figures denote plasma duration (s) or AIBN decomposition time (h) at 50 °C: DPPH, 10^{-4} mol/l; AIBN, 10^{-3} mol/l; Plasma, 100 W

progressive spectral changes of DPPH in DMSO with plasma duration are shown in Figure 26a. A similar spectral change was also observed when AIBN was thermally decomposed in the same solvent to form 2-cyano-2-propyl radicals in the presence of DPPH (Fig. 26b). However, the rate of spectral change by the plasma exposure was higher than that observed for AIBN decomposition. Linear decreases in the absorbance at 523 nm were obtained in both cases; the rate of DPPH consumption was calculated as 3.3×10^{-5} mol l^{-1} s^{-1} per 1-cm^2 surface of the solution, and that of AIBN decomposition at 50 °C as 2.5×10^{-9} mol l^{-1} s^{-1} for 10^{-3} mol l^{-1}, which is consistent with the literature. Since the rate of consumption of electrons formed by the plasma is proportional to the scavenger concentration, one can roughly estimate that the concentration of electrons generated in the plasma at 100 W and diffused successively through the liquid surface of 1 cm^2 is about 13,000 times higher than that of radicals formed by the decomposition of 10^{-3} mol l^{-1} of AIBN in DMF at 50 °C. This result indicates that an enormous number of electrons are formed in a very short period of plasma exposure, which then diffuse readily through the surface, reacting with viologen or DPPH [192].

Electron-accepting compounds such as iodine, potassium dichromate, potassium ferricyanide, and benzoquinone were successfully reduced in DMF by the plasma [193]. The reduction was considered to occur by the direct addition of high-energy electrons in the plasma to the inorganic substances to give I_3, Cr^0, and $K^4(Fe(CN)_6)$, respectively.

The novel reduction method using plasma seems to have potential applications to a variety of organic and inorganic compounds.

6 Surface Modification by Plasma Treatment

The effect of non-polymer-forming plasma (e.g., plasma of inorganic gases such as helium, argon, nitrogen, etc.) on a polymer can be viewed as the following two reactions [195]: 1) reaction of active species with the polymer, and 2) formation of free radicals in the polymer. The incorporation of nitrogen or oxygen into the polymer surface by N_2 or O_2 pasma are typical examples of the first effect. The latter effect generally leads to the formation of carbonyls and hydroxyls, and to some degree crosslinking depending on the type of substrate; however, the degradation of polymer at the surface leading to weight loss occurs in nearly all cases when polymers are exposed to plasma for a prolonged period of time (plasma etching, plasma ablation). Dehydrogenation of the polymer brings about an unsaturation and eventually results in extensive crosslinking to form stable and tough networks on the skin layer of the polymer (CASING).

The interaction of plasma with a polymer surface may be represented by the schematic diagram of reactions shown in Fig. 8 [22, 194].

Using these fundamental reactions of plasma, many polymer properties such as optical reflection, adhesion, friction coefficient, surface energy (wettability and water repellancy), permeability, and biocompatibility of conventional polymers can be controlled by the appropriate application of a plasma treatment. A typical industrial and large-scale application of this technique is the surface treatment of automobile bumpers. The advantage of this technique is the fact that plasma treatment usually changes the surface properties of the polymer without interfering with the bulk properties, simply because of the rather low range of its penetration. Wrobel [195] studied the plasma treatment of poly(ethylene terephthalate) fibers using various inorganic gases, and found that significant changes in surface structure as well as in the wettability of the fibers occurred, depending on the type of gas and the treatment conditions. The electron scanning micrographs of poly(ethylene terephthalate) fibers treated by oxygen plasma have also been investigated [196]. As far as natural fibers are concerned, Ward et al. [197] investigated plasma modification of cellulose and its derivatives and observed some changes of the elemental composition at the surface.

Yasuda et al. [198–200] studied the effect of plasma treatment on different fibers and fabrics. They used four nonpolymerizing gases: helium, air, nitrogen, and tetrafluoromethane. It was found that in some cases the etching of the fiber was accompanied by the implantation of the specific atoms into its surface. The model studies performed with nylon 6 have shown that plasma treatment, similar to plasma polymerization, may be carried out in the power-deficient range as well as in the gas-deficient range.

The effects of plasma treatment on weight loss by non-polymer-forming gas plasma, as well as plasma polymerization of hydrocarbons, on the surface of plasticizer-containing poly(vinyl chloride) (PVC) sheets, were examined. Gases used for the plasma treatment were argon, hydrogen, nitrogen, and ammonia. Hydrocarbons used for the deposition of plasma polymers were methane, ethylene, and acetylene. The efficiency of the prevention of plasticizer volatilization was evaluated by measurements of weight loss on heating and on extraction of plasticizer by hexane. It was found that any kind of plasma treatment or polymerization improves the prevention of volatilization of plasticizer. It was also found that there exists an optimum range of operational parameters given by (W/FM)t, where W/FM is the energy input per unit mass of gas or monomer, and t is the treatment time.

A decrease in the wettability of many hydrophilic fabrics has also been observed by using tetrafluoromethane plasma treatment. Okazaki et al. [201] have reported that any desired surface wettability of a plastic surface can be realized by changing

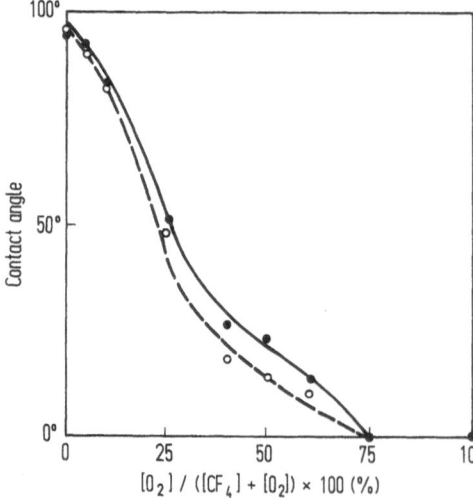

Fig. 27. The relation of O_2 concentration in O_2–CF_4 gas mixture and contact angle of poly (ether sulfone) films [202]

the relative concentration of the plasma gas, which was a mixture of oxygen and fluorine-containing gas (Fig. 27). Oxygen or air plasma treatment has been used to obtain a wettable surface on plastic films for printing or adhesion. A mixture of O_2–CF_4 as a plasma gas to control the plastic surface wettability was examined. Use of a third electrode consisting of a metal mesh could significantly decrease the etching process since this porous electrode effectively trapped the active ions. Thus, the authors were able to discuss the effect of O_2–CF_4 plasma treatment independently of the surface roughness of the plastic films.

Polytetrafluoroethylene (PTFE) and fluorinated graphite ($(CF)_n$ or $(CF_2)_n$ are very hydrophobic materials, the contact angle of a water drop having values of 110° for PTFE or 145° [202] for $(CF)_n$ and $(CF_2)_n$. Figure 28 shows that various plastic and graphite surfaces easily change to more hydrophobic surfaces by 1 min plasma treatment of CF_4 (i.e., the contact angle increased). The contact angle of $(CF)_n$ or $(C_2F)_n$ had a value of 160°, larger than Watanabe's [202] value of 145°.

The decreased wettability of the hydrophilic polymer film acquired by exposure to a CF_4 plasma discharge is reversed by aging. This is at present a serious problem which prevents the wide application of plasma treatment. The most probable reason is considered to be the "overturn" of polar groups on the film [203]. In order to investigate this overturn process more quantitatively, Y. Takaoka et al. [204], using ESCA, studied the elementary composition of the film treated with oxygen or nitrogen plasma. They found that the generated polar groups move to the inside of the film, and that hydrocarbon segments become exposed at the surface upon aging. The polar groups were difficult to overturn at low temperature (-196 °C), but this was easy at high temperature (100 °C). The overturn depends on the plasma discharge time, that is, the relaxation time for the surface energy of the films which were exposed to the plasma, becomes long. Immediately after plasma exposure, the polar groups with good wettability are mostly overturned, and even a long period of time after the exposure, the polar groups with weak wettability remain at the surface of the film.

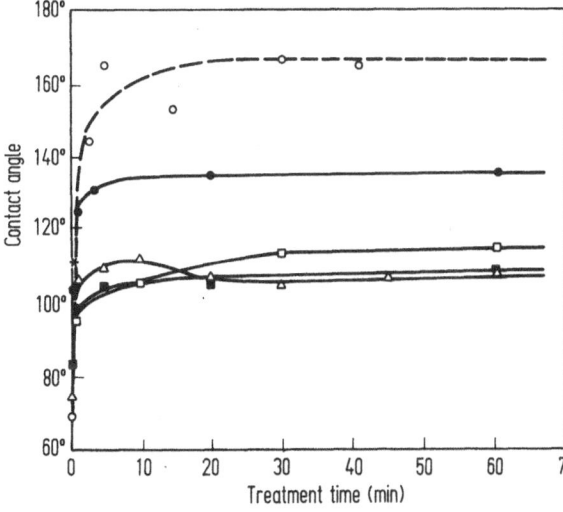

Fig. 28. The relation of plasma treatment time and contact angle; rf power: 50 W; CF_4 pressure: 3 torr for polymer, 5 torr for graphite [202]
O: graphite, △: polyimide, □: polyethyleneterephthalate, ●: polypropylene, ■: polyetherimide, *: polytetrafluoroethylene

Plasma etching can be applied to the preparative processing of polymeric specimen for electron microscopy. The effects of plasma etching can reveal new, pure surfaces of the sample and provide a thinner specimen which gives rise to a better resolution and contrast. Thus, techniques of sublimation and subsequent etching in a plasma of an electrodeless high-frequency discharge in conjunction with the method of vacuum-deposited carbon replicas, allowed to solve the fundamental problems concerning the morphology of commercial polymeric materials in the solid state.

A systematic investigation of the morphology was made by Bezruk [205–207]. For example, he carried out the initial evaporation of a polymer surface layer about 0.5–1 μm thick in a plasma gas-discharge unit at a rate of 100 nm s^{-1}. Oxygen or an inert gas served as the operating medium, the pressure being of the order of 5–2.5×10^{-1} torr. For the conditions used the mean electron energy in the plasma was about 5 eV, and the concentration of electrons was of the order of 10^{16} cm^{-2}. In this case

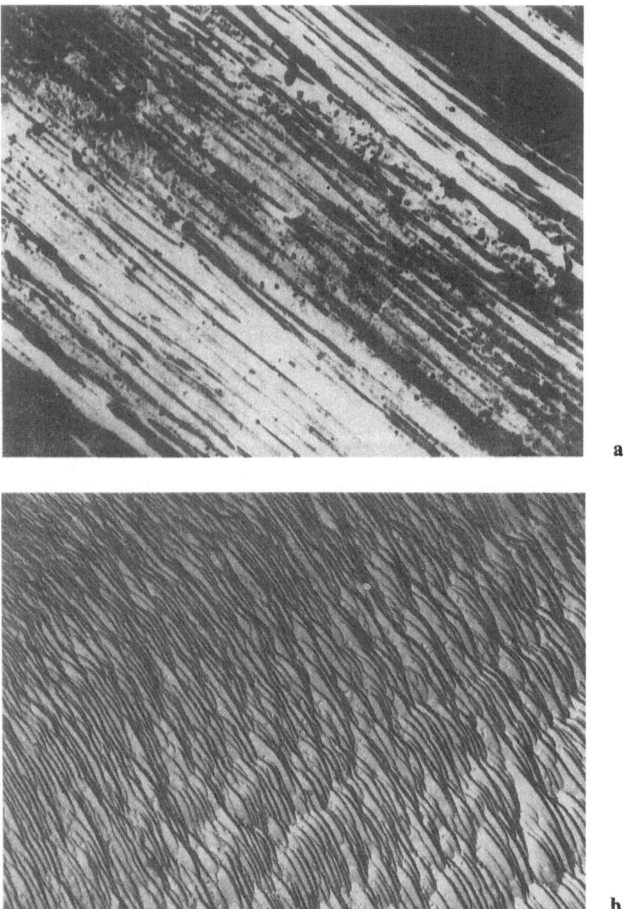

a

b

Fig. 29a, b. Microtissues on the surface of an unetched polycaproamide fiber. **b.** Morphology of the same fiber as in **a**, but at a depth of about 1 μm from the surface

the degradation rate of the crystalline and amorphous regions in the polymer was about the same, and the process consisted essentially in thinning of the sample without appreciable change of relief structure.

The morphological picture of a polycaproamide fiber surface prior to etching is shown in Figure 29 a. One might conclude that the fiber has a fibrillar structure. However, the morphology of the same fiber after etching is strikingly different. In place of fibrils an array of lamellar crystals is observed, the direction of which is at right angles to the orientation axis (Fig. 29 b). The patterns from low-angle X-ray diffraction and of Fraunhoffer diffraction were nearly identical both qualitatively (reflections on the meridian) and quantitatively (value of the long period).

7 Conclusions

Theoretical and experimental treatments of plasma chemistry processes on polymers have been concisely described. Main emphasis has been on properties of the obtained films and modified polymer surfaces and especially on their use in practice. In the future further applications are expected when better understanding of the plasma volume and surface processes is reached. This may allow to tailor respective films or surfaces according to the need of the particular application.

8 References

1. Hollahan JR, Bell AT (1974) Techniques and applications of plasma chemistry, Wiley, New York
2. McTaggart FK (1967) Plasma chemistry in electrical discharges, Elsevier, Amsterdam
3. Yasuda H (1985) Plasma polymerization, Academic Press, New York, London
4. Osada Y (ed) (1986) Plasma polymerization (in Japanese), Kagaku Dojin, Tokyo
5. Tien PK, Smolinsky G, Martin RG (1972) Appl Opt 11: 637
6. Spedding PL (1969) Chem Eng (London), 225: CF17
7. Venugoplan M (1971) React. Plasma Cond. 1: 1 Wiley Interscience, New York
8. Blaustein DD (ed) (1969) Chemical reactions in electrical discharges Adv Chem Series 80, Am Chem Soc, Washington DC
9. Wightman JP (1974) Proc. IEEE 62: 4
10. Nasser E (1971) Fundamentals of gaseous ionization and plasma electronics. Wiley Interscience, New York
11. Vossen JL, Cuomo JJ (1978) In: Vossen JL, Kern W (eds) Thin film processes, Academic Press, New York, London, p 11
12. Bell AT (1980) Springer, Berlin, Heidelberg, New York in: Veprek S, Venugopalan M (eds) Plasma chemistry, Topics in current chemistry, 94
13. Chen FF (1974) Introduction to plasma physics. Plenum Press, New York
14. Thornton JA (1982) in: Bunshah RF (ed) Deposition technologies for films and coatings. Noyes Publ. (USA), p 19
15. Butler HS, Kino GS (1963) Phys Fluids 6: 1346
16. Koenig HR, Maissel LI (1970) IBM J Res Dev 14: 168
17. Christophorou LG (1971) Atomic and radiation physics, Wiley Interscience, New York
18. Winters HF, Coburn JW, Kay E (1978) J Appl Phys 48: 4973
19. Coburn JW, Winters HF (1979) J Vac Sci Technol 16: 392
20. Taggart FK (1967) Plasma chemistry in electrical discharges, Elsevier, New York
21. Bell AT, Shen M (ed) (1979) Plasma polymerization, ACS Symposium Series 108, American Chemical Society
22. Yasuda H (1981) Macromolecular Reviews 16: 199

23. Williams T, Hayes MW (1966) Nature 209: 769
24. Carchano H (1974) J Chem Phys 61: 3634
25. Denaro AR, Ovens AP, Crawshaw A (1976) Europ Polym J 12: 205
26. Poll HV, Arts M, Wickleder KH (1976) Europ Polym J 12: 505
27. Thompson LF, Mayhan KG (1972) J Appl Polym Sci 16: 2317
28. Morita S, Bell AT, Shen M (1979) J Polym Sci, Polym Chem Ed 17: 2775
29. Suhr H (1968) A Naturforsch 23b: 1559
30. Neiswender DD (1969) Adv Chem Ser 80: 338
31. Duval M, Theoret A (1973) J Appl Polym Sci 17: 527
32. Duval M, Theoret A (1985) J Electrochem Soc 122: 581
33. d'Agostino R, Cramarossa F, De Benedicts S (1984) Plasma Chem Plasma Processing 4: 21
34. d'Agostino R, Capezzuto P, Bruno G, Cramarossa F (1985) Pure Appl Chem 57: 1287
35. Biederman H, Martinu L, Zemek J (1985) Vacuum 35: 447
36. Kay E (1986) Z Phys D-Atoms, Molecules, and Clusters 3: 251
37. Nyaiesh AR, Holland L (1984) Vacuum 34: 519
38. Holland L (1976) British Patent Application 1. 582: 231
39. Smith FW (1984) J Appl Phys 55: 764
40. Weissmantel C (1983) Proc IX Int Vac Congr, Madrid, 299
41. Wagner J, Wild CH, Pohl F, Koidl P (1986) Appl Phys Lett 48: 106
42. Tsuda M, Nakajima M, Oikawa S (1986) J Am Chem Soc 108: 5780
43. Saito Y, Matsuda S, Nogita S (1986) J Mat Sci Lett 5: 565
44. Angus JC, Koidl P, Domitz (1986) In: Mort J, Jansen F (eds) Plasma deposition of thin films. CRC Press
45. Biederman H (1986) Vacuum 37: 367
46. Biederman H, Martinu L, Slavinska D, Chudacek I (1988) Pure & Appl Chem 60: 607
47. Harding GL, Craig S (1979) J Vac Sci Technol 16 (3): 857
48. Craig S, Harding GL (1983) Thin Sol Films 101: 97
49. Window B, Savides N (1986) J Vac Sci Technol A4 (3): 453
50. Biederman H, Howson RP, McCabe I (1987) In: Proc IPAT '87, Brighton 152
51. Tibbitt JM, Bell AT, Shen M (1977) J Macromol Sci Chem 11: 139
52. Hozumi H, Kitamura K, Hashimoto H, Hamaoka T, Fujisawa H, Ishizawa T (1983) J Appl Poly Sci 28: 1651
53. Tajima I, Yamamoto M (1985) J Polym Sci Poly Chem Ed 23: 615
54. Ohno K, Ishii N, Sohma J (1983) Jap J Appl Phys 22: 996
55. Hoffman AS (1984) Adv Polym Sci 57: 142
56. Yasuda H, Baumgartner MO, Marsh HC, Yamanashi BS, Devito DP, Wolbarshy ML, Reed JW, Bessler M, Landers MB, Hercules DM, Carver J (1975) J Biomed Mater Res 9: 629
57. U.S.P. 338363 (1968; inv.: Menashi WP; U.S.P. 3701 628 (1972), invs.: Belmont LEA, Menashi WP
58. Yasuda H, Baumgartner MO, Morosoff N (1973) Annual Report NIH-NHL-73-2913, Research Triangle Institute, North Carolina
59. Kronick P (1975) Polym Prepr, Am Chem Soc, Div Polym Chem 16 (2): 441
60. Yamashita I (1975) Hyomen 17: 776
61. Chawla AS (1979) Trans Am Soc Intern Organs 25: 287
62. Takase M, Osada Y, Iriyama Y, Ishikawa Y, Sasakawa S (1985) Makromol Chem, Rapid Commun, 6: 495
63. Shen M (ed) (1976) Plasma chemistry of polymers, Dekker, New York
64. Buck KR, Davar VK (1970) Br Polym J 2: 238
65. Yasuda H, Lamaze CE (1973) J Appl Polym Sci 17: 201
66. Yasuda H, Lamaze CE (1971) J Appl Polym Sci 15: 2277
67. Yasuda H, Lamaze CE (1973) J Appl Polym Sci 17: 1719, 1533
67. Yasuda H, Lamaze CE (1973) J Appl Polym Sci 17: 1519, 1533
68. Hollahan JR, Wydeven T (1973) Science 179: 500
69. Terada I, Haraguchi T, Kajiyama T (1986) Poly J 18 (7): 529
70. Sakata J, Yamamoto M (1988) J Appl Polym Sci, Appl Polym Symp 42: 339
71. Osada Y, Takase M (1985) J Poly Sci, Poly Chem Ed 23: 2425–2439
72. Kitade T, Hozumi K, Kitamura K (1983) Bunseki Kagaku (Analytical Chemistry) 32: 368

73. Osada Y, Ohta F, Takase M, Kurimura Y, Mizumoto A (1986) J Chem Soc Japan 866
74. Osada Y, Ohta F, Takase M, Kurimura Y, Mizumoto A (1986) J Chem Soc Japan 873
75. Shinkai S, Ishikawa M, Manabe O, Mizumoto A, Osada Y (1985) Chem Lett 1029
76. Wydeven T (1977) Appl Optics 16: 717
77. Wydeven T, Johnson CC (1981) Polym Eng Sci 21: 650
78. Lerner NR, Wydeven T (1989) J Appl Polym Sci, in press
79. Bradley A (1965) Trans Faraday Soc 61: 773
80. Pliskin WA (1977) J Vac Sci Technol 14: 1064
81. Ristow D (1977) J Mat Sci 12: 1411
82. Yamada M, Hattori S, Morita S (1982) Jpn J Appl Phys 21: 10
83. Takai Y, Hayase Y, Mizutani T, Ieda M (1984) J Phys D 17: 399
84. Guastavino J, Bui HC (1975) Thin Solid Films 27: 225
85. Pender LF, Fleming RJ (1975) J Appl. Phys 46: 3426
86. Magno R, Adler JG (1977) Thin Solid Films 42: 237
87. Park YH, Tsutsumi H, Tasaka S, Miyata S (1986) Polym J 18: 713
88. Osada Y, Yu QS, Yasunaga H, Wang FS, Chen J (1988) J Appl Phys 64 (3): 1476
89. Osada Y, Yu QS, Yasunaga H, Kagami Y, Wang FS, Chen J (1989) J Polym Sci, Chem Ed, (1989) 27: 3799
90. Kay E, Dilks A, Hetzler U (1978) J Macromol Sci Chem A12: 1393
91. Kay E, Dilks A (1981) J Vac Sci Technol 18: 1
92. Kay E, Hecq M (1984) J Appl Phys 55: 370
93. Janca J, Pavelka P (1985) Scripta Fac Sci Nat Purk Brun 15: 261
94. Dimigen H, Hubsch H, Schaal V (1987) In: Proc of IPAT '87, Brighton, 207
95. Beale HA (1981) Industrial Res and Develop 23: 135
96. Martinu L, Biederman H, Zemek J (1985) Vacuum 35: 171
97. Morita S, Hattori S (1985) Pure & Appl Chem 57: 1277
98. Vossen JL (1979) J Electrochem Soc 126: 319
99. Holland L, Laurenson L, Hurley RE, Williams K (1973) Nucl Instr and Meth 111: 555
100. Hori M, Yoneda T, Yamada H, Morita S, Hattori S (1987) Plasma Chem and Plasma Process 7:
101. Itoh S, Yamada H, Morita S, Hattori S (1987) In: Akashi K, Kinbara A (eds) Proc. of ISPC 8, Tokyo p 1353
102. Asano Y (1983) Thin Sol Films 105: 1
103. Suzuki T, Ishimura K, Yoshida Y, Kashiwagi K, Murayama Y (1987) In: Akashi K, Kinbara A (eds) Proc of ISPC 8, Tokyo, p 428
104. Weissmantel C (1983) In: Proc of ISIAT '83 and IPAT '83, Kyoto, p 1257
105. Tkachuk BV, Marussi NY, Laurs EP (1973) Vysokomol Soedin, Ser A15: 2046
106. Sadhir RK, James WJ, Auerbach RA Thin Sol Films 97: 17
107. Kny L, Levenson LL, James WJ, Auerbach RA (1979) Thin Sol Films 64: 395
108. Kny E, Levenson LL, James WJ, Auerbach RA (1979) J Vac Sci Technol 16: 359
109. Kny E, Levenson LL, James WJ, Auerbach RA (1980) J Phys Chem 84: 1635
110. Inagaki N, Nishino T, Katsuura K (1980) J Polym Lettrs Ed 18 18: 765
111. Phadke SD (1978) Thin Sol Films 48: 319
112. Morosoff N, Patel DL, White AR, Umaria M, Crumbliss AL, Lugg PS, Brown DB (1984) Thin Sol Films 117: 33
113. Oehr CH, Suhr H (1987) Thin Sol Films 155:
114. Suhr H, Etspuler A, Feuer E, Oehr C (1988) Plasma Chem Plasma Process 8; 9
115. Osada Y, Mizumoto A (1986) J Appl Phys 59: 1776
116. Osada Y, Mizumoto A, Tsurata H (1987) J Macromol Sci-chem A24: 403
117. Osada Y, Yamada K (1987) Thin Solid Films 151(1): 71
118. Osada Y, Yamada K, Yoshizawa K (1987) Kobunshi Ronbunshu 44 (4): 267, 275
119. Yoshida Y, Kashiwagi K, Murayama (1987) J Chem Soc Japan 2013
120. Biederman H, Holland L (1983) Nucl Instr and Meth 219: 497
121. Biederman H, Ojha SM, Holland L (1977) Thin Sol Films 41: 329
122. Lehman HW, Frick K, Widmer R, Vossen JL, James E (1978) Thin Sol Films 52: 231
123. Roy RA, Messier R, Krishnaswami SV (1983) Thin Sol Films 109: 27
124. Craig S, Harding GL (1982) Thin Sol Films 97: 345
125. Sikkens M (1982) Solar Energy Mat 6: 403, 415

125. Sikkens M (1982) Solar Energy Mat 6: 403, 415
126. Weissmantel C, Ackermann E, Bewilogua K, Hecht G, Kupfer H, Rau B (1986) Vac Sci Technol 4: 2892
127. Biederman H, Howson RP, Kohoutek K, Chmel Z, Stary V (1989) to be published in Vacuum
128. Hogarth C, Iqbal T (1981) Phys Stat Sol A65: 12
129. Boonthanom, White M (1974) Thin Sol Films 24: 295
130. Luff PP, White M (1968) Vacuum 18: 437
131. Biederman H (1984) Vacuum 34: 405
132. Biederman H, Hon Jon Chjok, Martinu L, David J, Kadlec S, Lukac P (1987) In: Proc EMRS Conf, Strasbourg, C-1.8
133. Martinu L, Spatenka P, Biederman H, Sicha M (1986) Thin Sol Films 141: L83
134. Despax B, Flouttard JL (1986) Thin Sol Films 145: 233
135. Kay E, Hecq M, (1984) J Appl Phys 55: 370
136. Perrin J, Despax B, Kay E (1985) Phys Rew B32: 719
137. Perrin J, Hanchett V, Despax B, Kay E (1986) J Vac Sci Technol A4: 46
138. Martinu L, Biederman H (1985) J Vac Sci Technol A3: 2639
139. Martinu L (1986) Thin Sol Films 140: 307
140. Martinu L (1986) Solar Energy Mater 15: 21
141. Biederman H, Martinu L, Nespurek S (1987) In: Akashi K, Kinbara A (eds) Proc of ISPC 8, Tokyo, p 13964
142. Weissmantel C, Brener K, Winde B (1983) Thin Sol Films 100: 383
143. Biederman H, Chudacek I, Slavinska D, Martinu L, David J, Nespurek S (1988)(1989) Vacuum 39: 13
144. Abeles B, (1976) Appl Solid State Sci 6: 1
145. Biederman H, (1987) Materials Science XIII: 151
146. Wielonski RF, Beale HA (1981) Thin Sol Films 84: 425
147. Hecq M, Zieman P, Kay E (1983) J Vac Sci Technol A1: 364
148. Martinu L, Biederman H (1986) Vacuum 36: 477
149. Maxwell-Garnett JC (1904) Phil Trans R Soc London 203: 205, 237, 385
150. Martinu L, Biederman H, (1985) Plasma Chem and Plasma Process 5: 81
151. Martinu L (1987) Solar Energy Mater 15: 135
152. Bradley A, Hammes JP (1963) J Electrochem Soc 110: 15
153. Gazicky M, Yasuda H (1983) Plasma Chem and Plasma Process 3: 279
154. Johnes DI, Stewart AD (1982) Phil Mag B46: 423
155. Biederman H, Chmel Z, Fejfar A, Misina M, Pesicka V (1989) to be published in Vacuum
156. Carchano H, Lacoste R, Sequi Y (1971) Appl Phys Lett 19: 414; see Ref. 85
157. Biederman H, Lehmberg H, Pagnia H (1989) Vacuum 39: 27
158. Lehmberg H (1988) private communication
159. d'Agostino R (ed) (1989) Plasma deposition of polymer films. Academic Press New York, London (to be published)
160. Thurstans RE, Taylor AG, Oxley DP, Biederman H, Martinu L (1985) Vacuum 35: 219
161. Kay E (1987) In: Proc of EMRS Conf, Strasbourg, AII/III.5
162. Takeoka Y, Yasuda N (1983) In: Proc of ISIAT '83 and IPAT '83, Kyoto, p 993
163. Koeberle H, Memming R (1987) In: Proc of EMRS Conf Stasbourg, C-1.6
164. Grischke M, Brauer A, Thieme F, Memming R, Berndorf C (1987) In: Proc. of EMRS Conf, Strasbourg, C-1.7
165. Nishizawa J, Itoh K, Okuno Y, Sakurai F (1985) J Appl Phys 57: 2210
166. Mishima T, Takahashi K (1983) J Appl Phys 54: 2153
167. Mino N, Kobayashi M, Konagai M, Takahashi K (1986) J Appl Phys 59(6): 2216
168. Osada Y, Bell AT, Shen M (1978) J Polym Sci Polym Lett Ed 16: 309
169. Johnson D, Osada Y, Bell A, Shen M (1981) Macromolecules 14: 118
170. Osada Y, Bell A, Shen M (1979) In: Shen M, Bell AT (eds) Plasma polymerization, Am Chem Soc Symp Ser No. 108, Am Chem Soc, Washington DC 263
171. Osada Y, Shen M, Bell AT (1978) J Polym Sci Polym Lett Ed 16: 669
172. Odajima A, Nakase Y, Osada Y (1979) In: Shen M, Bell AT (eds) Plasma polymerization, Am Chem Soc Symp Ser No. 108, Am Chem Soc, Washington DC 263
173. Osada Y, Hashizume M, Tsuchida E, Bell AT (1981) Nature (London) 286: 593

174. Osada Y, Mizumoto A (1985) Macromolecules 18: 302
175. Osada Y, Takase M, Iriyama Y (1983) Polym J 15: 81
176. Paul CW, Bell AT, Soong DS (1985) Macromol 18: 2312
177. Paul CW, Bell AT, Soong DS (1985) Macromol 18: 2318
178. Kuzuya M, Kawaguchi T, Nakanishi M, Okuda T (1986) J Chem Soc, Faraday Trans 1, 82: 1441
179. Kuzuaya M, Kawaguchi T, Yanagihara Y, Okuda T (1986) J Polym Sci, Polym Chem Ed 24: 707
180. Osada Y, Iriyama Y, Takase M (1981) Kobunshi Ronbunshu 38: 629
181. Alfrey Jr T, Prince CC (1947) J Polym Sci 2: 101
182. Bamford CH Jenkins AD, Johnson R (1959) Trans Faraday Soc 55: 418
183. Flory PJ (1953) Principles of polymer chemistry, Cornell U.P., Ithaca, NY, Chap. 4
184. Osada Y, Takase M (1983) J Polym Sci, Polym Lett Ed 21: 643
185. Klein JA, Bell AT, Soong DS (1987) Macromol 20: 782
186. Osada Y, Iino Y, Iriyama Y (1982) Chem Lett 171
187. Osada Y, Iriyama Y, Takase M, Iino Y, Ohta M (1984) J Appl Polym Sci, Appl Polym Symp 38: 45
188. Osada Y, Takase M (1983) J Chem Soc Japan 439: 831
189. Osada Y, Iriyama Y (1984) Thin Solid Films 118: 197
190. Osada Y, Honda K, Ohta M (1986) J Membrane Sci 27: 327
191. Osada Y, Iriyama V (1982) J Am Chem Soc 104: 2925
192. Osada Y, Iriyama Y (1984) J Phys Chem 88: 5951
193. Osada Y, Honda K, Mizumoto A (1984) J Chem Soc Japan 1693
194. Yasuda H, Hsu T (1978) Surf Sci. 76: 232
195. Wrobel AM, Kryszewski M, Rakowski W, Okoniewski M, Kubacki Z (1978) Polymer 19: 908
196. Blakey PR, Alfy MO (1978) J Text Inst 69: 38
197. Ward TL, Jung HZ, Hinojosa O, Benerito RR (1979) J Appl Polym Sci 23: 1987
198. Yasuda T, Gazicki M, Yasuda H (1984) J Appl Polym Sci, Polym Symp 38: 201
199. Yasuda H, Lamaze CE (1973) J Appl Polym Sci 17: 1533
200. Gazicki M, Yasuda H (1983) Proc 18th ACS Meeting, Symp on Plasma Polymerization and Plasma Treatment, ACS, Washington DC, p 308
201. Moriwaki T, Okazaki S, Kogoma M, Kasai H, Takahashi K (1987) J Phys D Appl. Phys 20: 147
202. Watanabe N, Ashida Y, Nakajima T (1982) Bull Chem Soc Jpn 55: 3197
203. Yasuda H, Sharma AK, Yasuda T (1981) J Polym Sci Polym Phys Ed 19: 1285
204. Taru Y, Takaoka K (1986) Kobunshi Ronbunshu 43(6): 361
205. Bezruk LI, Lipatov YS (1972) J Polym Sci Part C 38: 337
206. Bezruk LI, Khoreva G (1982) Mechanics of Composite Materials (in Russian) 3: 387
207. Bezruk LI, Khoreva G (1982) Composite Polymer Materials 12: 39

Editor: G. and S. Olive
Received April 26, 1989

Flow of Polymerizing Liquids

A. Ya. Malkin
Research Institute of Plastics, USSR 111 112, Moscow E-112, Perovskii pr. 35,
P. V. Zhirkov
Academy of Sciences of USSR, Institute of Structural Macrokinetics, USSR 142 432
Moscow District, Chernogolovka

The present work is a review of a particular field of polymer science which is developing fast and is at the interface of rheology, chemical hydrodynamics, macrokinetics, and the kinetics of polymerization processes. It involves the analysis of the way the viscosity growth affects a chemical reaction, i.e. the hydrodynamic, thermal, and concentration fields during polymerization.

In this review, the rheokinetic approach is presented, complete mathematical and physical statements of the problem are given, and the operation of a tubular polymerization reactor is analyzed as an example. The fundamental necessity of using the rheokinetic approach, whenever there is a sharp growth in the viscosity, is demonstrated. The trends of further investigation are presented.

Advances in Polymer Science 95
© Springer-Verlag Berlin Heidelberg 1990

1 Introduction

Rheokinetics of the process of formation of linear and cured (network) polymers has two aspects. The first of them, which is complete, concerns the phenomenological description of the variation of rheological properties during polymerization or subsequent reactions. In this case, the final result is a set of constants which are characteristic for a given compound but dependent on its composition and temperature. The second aspect, for which the first approach is only an introduction, is an analysis of hydrodynamic situations and the solution of engineering problems for the media whose properties are described by the appropriate rheokinetic equations. Here the final objective is to find velocity and temperature fields, pressure-output dependence, dissipation of energy and so on. This second aspect of the problem is significant both for the rheology of the polymeric materials, which can be called "living", and for real production processes.

This review calls the reader's attention to the second aspect of the general rheokinetic problem, namely, the hydrodynamics of liquids with variable rheological properties and to more specific problem of polymerizing liquids, with the aim of finding out specific effects arising during the flow of such liquids.

The general approach to solving hydrodynamic problems on "rheokinetic" liquids is based on the representation of their rheological properties by the most simple phenomenological models. Here, the chemistry of the process is deliberately neglected and the idea of a "black box" is utilized, in which formal methods of describing experimental data are applied not only due to an insufficient understanding of the essence of the process, but mainly due to the tendency to ensure the possibility of obtaining explicit final results.

Hydrodynamics of "rheokinetic" liquids is a theoretical basis for many production processes, namely, RIM-process, processing of items from thermosetting resins, molding of elastomers, plastisols, etc. It is important to note that different physicochemical processes are superimposed [1–3], first of all heat exchange and flow [4], because rheokinetic phenomena are always accompanied by heat output and occur under nonisothermal conditions. Then, it is necessary to allow for the possibility of crystallization of the product taking place [5, 6], etc.

From the end of sixties, the principal studies in the theory of chemical technology were based on mathematical and physical modelling of the total set of superimposed processes. Vigorous development of computers and numerical methods of analysis promoted a fast development of investigations and continuous complication of models, which enable us to percive new details of the processes. At present, the fundamental physical principles and phenomena are understood in principle, mathematical models of processes have been developed in the main types of reactors, the fields of their application have been determined and computational methods for solution and analysis have been defined. Since the mid 1970s, the main attention of the researchers has been attracted to the study of the peculiarities of processes.

Rheokinetic processes for polymerizing media have clearly defined specific properties. These are as follows: a complex kinetic scheme, variation of molecular-mass distribution (MMD) of products with time, and viscosity growth during reaction. The first two peculiarities were investigated in detail in numerous works both in general form and in connection with different types of polymerization and polycondensation reactions. The results of studies along these lines, different approaches to the analysis and general principles are formulated and generalized in a number of monographs [e.g. 7, 8].

At the same time, the characteristic feature of the polymerization process — the viscosity growth — in polymerization reactors was not studied for a long time and the first paper published in this field was the work of Lynn and Huff [9]. This fact seems quite strange because everybody understood that the viscosity of a polymer was the most important property, and that its magnitude and growth affected substantially the process (for instance, a long time ago the assumption was made that the viscosity growth affected the rate of chain termination reaction).

In connection with chemical engineering of polymeric materials, flow processes were usualy considered from the point of view of processing of the materials already being prepared, (cf. e.g. the monograph [10]). This approach has led to the development of an extensive field called polymer rheology dealing with quantitative description of peculiar mechanical properties of polymeric materials [11] and also has stimulated the development of a general approach to the analysis of flow processes, i.e. to the formulation of a general system of moment equations, describing the flow processes of these rheologically complex liquids. This achievement has mode it possible to solve a set of boundary problems which model the principal flow patterns in processing equipment. The problems belonging to this field had a common feature — a flowing liquid was considered as a stable one; therefore, its rheological properties could be very complicated depending on thermodynamic parameters (temperature and pressure), but principally, they did not vary with time. Viscoelastic effects (i.e. time variation of the *behavior* but not *properties*) do not belong to the class of rheokinetic phenomena.

At present, there exists a field in chemical engineering of polymeric materials that requires theoretical substantiation. This is the flow of "rheokinetic" liquids, i.e. media with rheological properties varying with time. Lately, a number of papers have appeared in which general problems, were formulated and some specific problems were solved. At present, we can say that fundamental principles of the process are established, its physical picture is understood and new phenomena are being found. Along with this, a number of problems have developed at a simple model level and many practically important and theoretically interesting problems remain unsolved. The aim of this review is to give a critical review of the present state of this field of science and to formulate perspectives for its further development.

2 Statement of the Problem

The time variation of rheological properties of a flowing liquid is a fundamentally new factor in the range of problems under consideration. First of all, the viscosity is concerned. It is precisely the viscosity that is the basic characteristic of the flowing medium. In addition, the whole combination of rheological properties of polymeric materials is connected with non-Newtonian polymer viscosity [11]. Finally, the magnitude of the viscosity reflects the molecular properties of the polymer. Variation, usually an intense growth of viscosity, reflects chemical processes in the flowing material which results in the increase in length and concentration of macromolecules in the flowing reaction medium.

The variation of molecular composition of a liquid is described by a kinetic scheme of polymerization; a complete scheme includes a great number of different elementary reactions. During analysis of macrokinetics of polymerization processes, i.e. the development of temperature, concentration and hydrodynamic fields in the course of the process, a separate consideration of each step is usually very difficult and ineffective. It is more rational to introduce a unified internal parameter, the conversion β, which characterizes the state of the material [12–14]. For simple kinetic polymerization schemes, the conversion corresponds to the change of concentration of reactive groups or to the conversion of a monomer.

The basic initial task consists of establishing the law of variation of β with time t, and a relation between β and viscosity. The behavior of $\beta(t)$ as a function of t is determined by a phenomenological equation for the kinetics of a specific process, which can be conveniently written in the following form [11]:

$$\dot{\beta} = kf(\beta) \exp \frac{-E}{RT}. \tag{1}$$

Here k is a pre-exponential factor, $f(\beta)$ is the kinetic function, E is the (apparent) activation energy of the process, R is the universal gas constant, T is temperature in Kelvin.

Such notation restricts the problem, first, by assuming that a single parameter β exists, which determines the behavior of the process (in a general case this may be an over simplification; indeed, during radical initiated polymerization, for example, the kinetic function depends on concentrations of both monomer and initiator, whose ratio varies with time in different ways according to what degree the process is non-isothermal). Second, by postulating that the effect on the reaction rate $\dot{\beta}$ on conversion β and temperature T can be separated, i.e. by considering that the constants entering the kinetic function $f(\beta)$ are independent of temperature and the activation energy E of a chemical process is independent of conversion. These restrictions are sufficient above the glass transition temperature in the majority of cases. There are cases when these postulates are not fulfilled and it is necessary, for instance, to take into account E as a function of β. However, the indicated restrictions are not fundamental. If these approximations are not applied, the solution and analysis of the problem become complicated, but the basic approach remains unchanged. In the final analysis, for determining the hydrodynamic fields during the flow of a liquid with increasing viscosity only the relation between the viscosity and the residence time in the channel is important. Moreover, this relation can be given in any empirical form for the numerical solution.

The form of the $f(\beta)$ function as well as numerical values of the constants, i.e. the preexponent k and activation energy E, depend on the mechanism of polymer formation and on the nature of the reactants. Some examples are discussed below.

The viscosity relation $\eta(\beta)$ and $\eta(t)$ are also determined by the mechanism of the reaction, composition of the system and temperature [12, 13]. There are no universal formulas for this case though the literature given certain general laws [14]. Thus, it is usually assumed that there exists an exponential dependence (via the Arrhenius formula) of the viscosity on temperature $\eta = \eta_1 \exp(U/RT)$ and power dependence of the viscosity on the molecular weight: $\eta \sim M^{3.4}$, beginning from a certain value of M or β, $M(\beta = \beta_1)$, $\eta \sim M$ up to this value. In the literature, the $\eta(\beta)$ dependence takes different forms which more or less simplify its real form, namely:

(1) stepwise: $\eta = \begin{cases} \eta_0 & t < t^* \\ \eta_f & t \geq t^*, \end{cases}$

where it is possible that $\eta_f = \infty$;

(2) exponential: $\eta = \eta_0\, e^{AB}$;

(3) power: $\quad \eta = \eta_0 + A_1 \beta^n$;

(4) scaling: $\eta = \eta_0 \left(1 - \dfrac{\beta}{\beta^*}\right)^{-n}$;

but the last dependence has purely physical meaning and holds in the transition region. The dependences $\eta(T)$, $\eta(M)$ and $\eta(\beta)$ are given schematically in Fig. 1.

In addition, in model calculations it is assumed that the rule of "logarithmic additivity" is valied, i.e. all these dependences can be represented as a product of independent multipliers:

$$\eta = \eta_1 \exp \frac{U}{RT} M^{3.4} \beta^n , \tag{2}$$

where $n = 3$–7. However, the effect of the initial viscosity variation is omitted here. It may often be neglected because the viscosity of initial reactants is small in comparison with the viscosity in the main stage of the process. Therefore, formula (2) is sufficiently accurate for model calculations of the production process.

Going over to the description of any concrete process, different corrections, refinements, coefficients and empirical constants appear related to real properties of a solution or melt, chemical nature of monomers, properties of polymer chains and

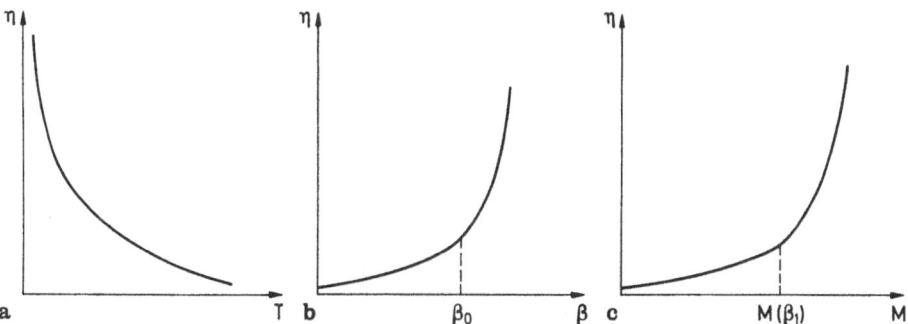

Fig. 1a–c. Schematic representation of the relationship between the viscosity η of polymerizing liquid and temperature T **(a)**, conversion β **(b)**, and molecular weight M **(c)**

many other factors. The literature dealing with the analysis of such dependences is generalized in particular in the monograph [14].

For quantitative calculations of real technological equipment, the $\eta(\beta)$ dependence must be known in a more accurate form and it is necessary to substitute the convenient analytical expressions and approximate data with sufficiently exact formulas. Since the appropriate calculations, for example, of flow in a tube reactor can be carried out only numerically by a computer, the form and representation of the dependence are insignificant. A simple tabulated version may turn out to be even preferable.

It is worth noting that there are two basically different situations of the viscosity variation during the polymerization process. First: the reative mass remains liquid, i.e. its ability to shear flow is maintained: $\eta(t \to \infty) < \eta_f$ until the reaction is completed ($\beta = 1$). Second: the reactive mass loses possibility to flow (gap) $\eta(\beta \to \beta^*) \to \infty$ at a certain critical conversion. $\beta = \beta^* < 1$. Both cases are of real technological interest; the corresponding curves are given in Fig. 2.

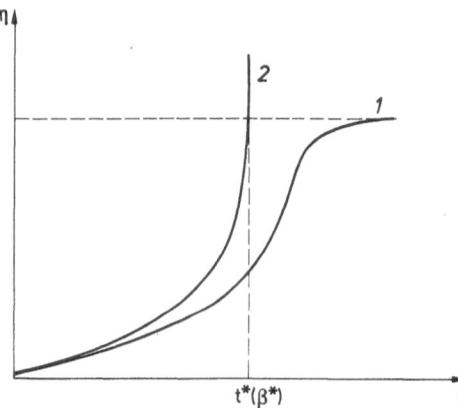

Fig. 2. Dependence of viscosity with conversion on time: 1 — keeping the ability to shear flow; 2 — losing the possibility to flow

In the first case, the relation between the viscosity and conversion can be modelled by any relation $\eta(\beta)$ given above: stepwise, exponential, power dependence. In the second case, this relation can be described, as a rough approximation, as a stepwise:

$$\eta = \begin{cases} \eta_0 & \text{at } t < t^* \\ \infty & \text{at } t \geq t, \end{cases} \tag{3}$$

For more accurate description, the following formulas are given in the literature:

$$\eta = \frac{\eta_0}{t^* - t} \quad (t < t^*) \quad \text{or} \quad \eta = \frac{\eta_0}{\beta^* - \beta} \, (\beta < \beta^*) \, .$$

A change in the reactive medium composition during the chemical process results in the change of the complex of its rheological properties. It is well known [11] that most polymeric materials (solutions and melts) show non-Newtonian properties

(pseudoplasticity, viscoelasticity, viscoplasticity, etc.). It is obvious that non-Newtonian behavior has an effect on the flow pattern of a polymerizing liquid, but very
little can be found in the literature.

Proceeding from the estimates of possible viscosity variations caused by non-Newtonian effects and rheokinetic phenomena, it is obvious that the effect of non-Newtonian factors on flow regularities always plays a minor role as a correction factor.
The case, when the non-Newtonian behavior affects qualitatively shear flow, for
example, if a reactant acquires the ability to slip along the wall of a channel, will be
discussed in detail later.

The physical pattern of motion of a polymerizing liquid in a channel depends
qualitatively on whether the reaction medium loses the ability to flow during the reaction. Therefore, in this review we discuss at first the problems of flow when $\eta < \infty$
for any t and later the effects of a loss of the ability to flow will be discussed.

Concerning the problem of the flow of polymerizing liquids in chemical engineering,
it is necessary to separate the effects of nonisothermality of the process which are
fundamental to the problem [1, 15]. There are three sources of nonisothermality
which can play different roles in real production processes: heat exchange with the
surrounding medium (thermostat), heat effect of a chemical reaction, and dissipative
heat output, i.e. heat output because of viscous dissipation at shear flow. The last
factor may be especially significant in the final stages of the process when the viscosity of the reactive mixture becomes very high. It is very significant and important
that a complicated nonlinear feedback exists between thermal factors, the course
of a chemical reaction and hydrodynamic situation in the channel of the reactor.
Thus, a local increase in heating, for example, leads to an acceleration of chemical
transformation and the viscosity of the flowing medium is affected and its velocity
is changed. There are rather many links in this logic chain, and the effect of a number
of factors is competitive (for example, the effect of temperature and conversion growth
on the viscosity).

All these factors yield the following heat balance equation for the flow of a polymerizing liquid under specific geometric conditions, particularly, in a cylindrical tube
(model of a tubular reactor):

$$\frac{\partial T}{\partial t} + V_r \frac{\partial T}{\partial r} + V_z \frac{\partial T}{\partial z} = \frac{\lambda}{C\varrho} \left(\frac{\partial^2 T}{\partial r^2} + \frac{1}{r} \frac{\partial T}{\partial r} + \frac{\partial^2 T}{\partial z^2} \right) +$$

$$+ \frac{qk}{C\varrho} f(\beta) e^{-E/RT} + \frac{1}{C\varrho} \eta(\beta, T, M) \left(\frac{\partial V_z}{\partial r} \right)^2. \quad (4)$$

Here r and z are the radial and axial coordinates; V_r, V_z are the radial and axial
components of the flow velocity; C, ϱ, λ are the heat capacity, density and heat
conduction of the liquid, which in a general case may depend on β, but usually this
dependence is weak if initial reactants are liquids; q is the thermal effect of the reaction.

By consistent arrangement of Eqs. (1) and (4) it is assumed that there is a unique
chemical process following Arrhenius law, i.e. a reaction which is described well
by Eq. (1) and apparent constants k, E and kinetic function $f(\beta)$. It has already been
mentioned that this assumption simplifies the model significantly but it is not necessary. In a general case Eq. (1) is replaced by a system of material balance equations

for all substances involved in the reaction and the second term of Eq. (4) represents the sum of heat output at all stages of the process.

A closed system of equations is obtained when dynamic equations are added to Eqs. (1)–(3)

$$\frac{\partial P}{\partial z} = \frac{1}{r} \frac{\partial}{\partial r} \left[\eta(\beta, \, T, \, M) \, r \, \frac{\partial V_z}{\partial r} \right] \tag{5}$$

$$\frac{1}{r} \frac{\partial}{\partial r} (rV_r) + \frac{\partial V_z}{\partial z} = 0$$

(P is the hydrostatic pressure). Let us discuss in detail the assumptions that were used for the derivation of Eq. (5). We assume that the process proceeds quasistationary from the point of view of hydrodynamics, i.e. the velocity field is quickly tuned according to the variation of temperature and concentration fields, which is valid for high values of the Prandtl Number, $\text{Pr} = \frac{C\eta}{\varrho}$, $\text{Pr} \gg \text{I}$. Real values of the Prandtl Number for polymeric systems in the beginning of polymerization are $\text{Pr} = 10 \div 100$ and they quickly increase proportionally to viscosity in the course of the reaction. In a certain conversion range, Pr is greater than $100 \div 300$. The general assumption is that the tube is very long and all quantities and their derivatives change along the length of the channel much more slowly than along its radius. We neglect inertial terms due to a smooth variation of the velocity profile from section to section. We think as well that the changes of the axial component of the velocity along the radius is much stronger than in the axial direction $\frac{\partial V_z}{\partial r} \gg \frac{\partial V_z}{\partial z}$ and also that the change of the radial component along the length of the tube is sufficiently smooth, $\frac{\partial V_r}{\partial z} \ll \frac{\partial V_z}{\partial r}$, which follows from the previous assumptions. In addition, we neglect the pressure variation along the radius of the tube.

In the statement of the rheokinetic problem expressed by Eqs. (1), (2), (14) and (5), the inclusion of the viscosity variation during the process and the effect of this variation on the flow velocity distribution is important and new; the consideration of a complicated hydrodynamic process as a primary stable flow with fixed viscosity profile (plug-like or parabolic) whose shape is independent of any external or internal conditions, has been abandoned.

3 Flow Without Liquid-to-Solid Transition

3.1 Stirred Tank Reactors

Rheokinetic investigations of this kind are aimed at revealing hydrodynamic regularities of the operation of reactors. It should be noted that thermal aspects of chemical reactions in different types of reactors have been studied thoroughly and in detail, see, for example, monographs [4, 16–19]. The problems of the existence and multiplicity of steady-state solutions, non-stationary development of the process including oscillating conditions and stability problems have also been studied. Critical conditions and the values of critical factors have been found. Solving the whole complex of these problems makes it possible to formulate the general theory of reactors.

The greater part of investigation in this field is carried out on the assumption of the simplest hydrodynamic behavior of a medium: (a) completely stationary medium; (b) a medium whose rheological properties are fully identical at every point of the volume; (c) flow of reactants with constant velocity at each point of the reactor. The last two idealized models of hydrodynamics in the theory of reactors received the names "stirred tank reactor" (STR) and "plug-flow reactor" (PFR). These models are very convenient since they simplify considerably the hydrodynamic part of the problem (of the type of Eqs. (5)). In addition, they allow us to determine the general mechanisms of the process at various levels of approximation. In a number of cases, the models describe quite adequately the hydrodynamics of the process. The PFR model is applicable to the flow in a tubular or other analogous reactors under turbulent conditions, i.e. to gas flow. STR describes sufficiently well the process in a reactor with intensive agitation of a low-viscosity medium.

When dealing with polymerization processes, where low-viscosity and high-viscosity products are simultaneously mixed, the application of the PFR and STR models becomes extremely doubtful.

We believe that the main problem of the theory for the stirred reactors is to estimate the residence time distribution (RTD) over the volume of the reactor. It should be said that at present papers in this field and even attempts to develop an approach to a description of such problem are absent. This applies to mixing devices for mixing inert liquids. In addition, it is evident that one can distinguish more or less clearly stagnancy zones with a relatively high residence time and zones of relative short-circuiting of the stirred liquid for almost any stirred reactor constructed as a mixing device. The construction of a reactor with "successful" hydrodynamics depends on the engineers' intuition, previous experience and empirical investigations including the creation of response functions by the method of a "black box". All this does not give any quantitative estimates of the flow pattern.

A sharp increase of viscosity aggravates the situation significantly. Due to a positive feedback "residence time ↔ viscosity", the viscosity growth during polymerization leads to slowing down of the flow in the stagnancy zone, i.e. to an increase of residence time and, as a result, to the viscosity growth, and so on. Small initial differences in residence time increase greatly and turn into qualitative peculiarities of the behavior of rheokinetic liquids in different zones of the reactor. Lamination effects (stratified flow) are possible. Obviously, the results of such investigations would be interesting from many points of view. However, it is difficult to suggest a satisfactory quantitative approach to such a problem.

The attempts to model the hydrodynamics of mixing devices were made only in Refs. [20] and [21] concerning the estimate of the role of viscoelastic effects. In Ref. [20], the stream-lines for a model agitator in the form of a sphere are studied the flow patterns are defined: (α) typical for viscous liquid, (β) transient, (γ) typical for elastic liquid. The most important result of Ref. [20] is the finding that the transition from one type of the flow to another is defined by the ratio of Weissenberg and Reynolds Numbers:

$$m = \frac{We}{Re} = \frac{\theta/N}{\varrho R_0^2 N/\eta} = \frac{\theta\eta}{\varrho R_0^2 N^2}$$

(R_0 is the radius of the sphere, N is the angular velocity, θ is the relaxation time). In Ref. [21], this approach is combined with the experimental investigation of correlations between the parameters characterizing viscoelastic properties of polymer solutions and peculiarities of the flow (stream-line behavior) induced by the agitator.

The effect of viscosity growth on the thermal pattern of polymerization in a stirred tank reactor is analyzed in Ref. [22]. The authors assumed the STR model but introduced empirical dependences (taken from the literature) relatively to the coefficient of heat transfer through the reactor's wall as a function of viscosity, $\alpha(\eta)$. Deterioration of heat exchange with the surrounding medium due to the viscosity growth

$$\alpha = \alpha_0 \eta^{-n} \qquad (n = 0.3 \div 0.4)$$

(α_0 is the initial value of the heat transfer coefficient) results in the appearance of peculiar thermal behavior (Fig. 3).

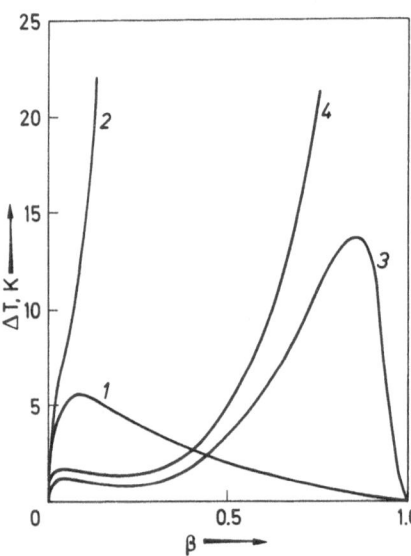

Fig. 3. Temperature curves: $1, 2$ — hear exchange with the surrounding medium is independent of the viscosity; $3, 4 - \alpha = \alpha_0 \eta^{-n}$ $1, 3$ — low-temperature process; $2, 4$ — high-temperature process

It is to be recalled that for sufficiently high values of the apparent activation energy of the chemical process E and thermal effect of the reaction q (more exactly, for

$$\frac{RT_0}{E} \ll 1 \quad \text{and} \quad \frac{C_\varrho RT_0^2}{qc_0 E} \ll 1)$$

the process proceeds in two qualitatively different stages; low temperature and high-temperature states (lines 1 and 2 in Fig. 3). In this case, the transition from one type of behavior to the other one occurr in a narrow range of variation of the value of Semyonov Number $\varkappa$ = (heat output intensity)/(heat removal intensity),

$$\varkappa = \frac{E}{RT_0^2} \frac{qc_0 k \exp(-E/RT_0)}{\alpha S/V} = \varkappa_{cr}$$

(T_0 is the temperature of the surrounding medium, c_0 is the initial concentration, S and V are the surface area and volume of the reactor).

The viscosity growth and the corresponding decrease in heat removal may lead to thermal disbalance and to repeated temperature growth in the region of high conversions. Calculations and experiments have shown that the temperature growth amounts to 20 ÷ 30 K (Fig. 3, line 3) and even the transition to high-temperature condition (line 4) may occur.

3.2 Plug Reactors. Thermal Pattern

The same thermal regularities are valid for plug-flow reactors and for STR, which means that the process occurs in low-temperature conditions for $\varkappa < \varkappa_{cr}$ (Fig. 4, line 1) and in high-temperature conditions for $\varkappa > \varkappa_{cr}$ (Fig. 4, lines 2, 3). There are

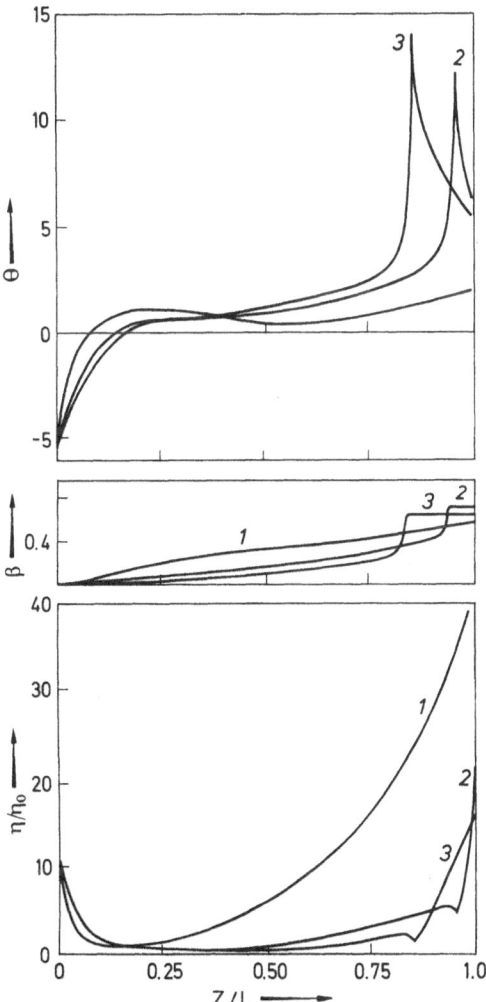

Fig. 4. Profiles of temperature, conversion of monomer and viscosity (dimensionless quantities) accross the reactor length (1 — low-temperature process; $2, 3$ — high-temperature process)

qualitative differences in these two states between longitudinal profiles of temperatures and concentrations which are averaged over the section, and this determines the
qualitative difference of the viscosity profiles.

The existence of a sharp boundary between high- and low-temperature conditions
for the polymerization process is shown experimentally in an example of styrene in
Ref. [23]. In the simplest PFR model, the coordinates of all points of the temperature
and concentration profiles are displaced in proportion to the flow rate, i.e. there exists
a linear dependence between the flow rate and the position of the temperature maximum.

The situation becomes complicated when a nonlinear feedback appears between the
heat output and temperature profile. Such feedback as applied to PFR may be heat
conduction in the axial direction (a large heat flow along metal walls of the reactor).
In this case, a multiplicity of steady states (steady-state solutions) may appear. This
solution means that to one value of the flow rate in a certain range of its variation
(possibly discontinuous [19]), correspond three steady-state solutions: low (low-temperature condition), high (high-temperature condition) and intermediate — absolutely
unsteady [19, 23]. The reactor begins to operate under specific regime according to the
initial conditions. The reaction process in PFR is in complete analogy to the combustion process. Changing the flow rate, one can observe "ignition" (abrupt transition
from low- to high-temperature conditions) and "extinction" (break-down of high-
temperature conditions), moreover, these transitions are of a hysteresis nature. The
region of nonunique solution is determined by the values of the parameters and may
have a different form [24].

In STR of periodical action (the problem with initial conditions), the solution is
always unique and the process is unsteady by definition, while in STR of continuous
action (with a constant feed of reactants) there may exist a multiplicity of steady-state
solutions and the effects described above for PFR.

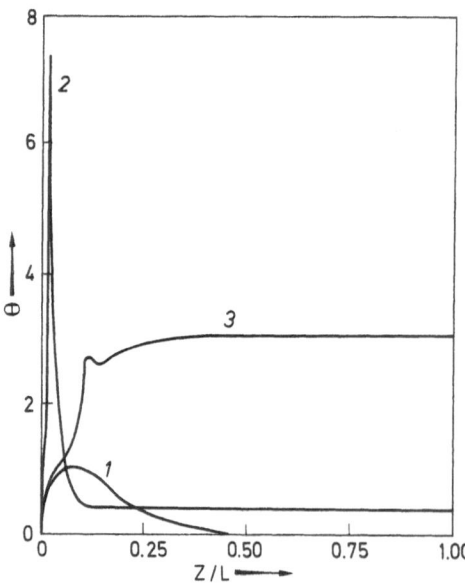

Fig. 5. Profiles of dimensionless temperature
in a screw reactor. *1* — low-temperature process; *2* — high-temperature process; *3* — inter-
mediate-temperature process

The heat output due to viscous friction (dissipative heat source; the intensity of this heat output $W_{mech} \sim \dot{\gamma}\eta(\beta, M, T)$ where $\dot{\gamma}$ is the shear rate) may cause a nonlinear variation of temperature profiles $T(z)$ with the variation of flow rate. The temperature behavior of the process is determined by the ratio of all thermal factors [25]. The estimates show that for tubular reactors the dissipative heat output is small even in the case of a strong viscosity growth during polymerization, but the heat output can play a role of a "trigger" for the transition of the process from low-temperature region to high-temperature region near critical conditions [26]. In the same way the dissipation of energy is always a minor factor in stirred reactors during agitation [22]. However, under specific conditions of intensive agitation in the channel of a screw polymerization reactor, the dissipative heat output is the determining factor and causes the appearance of new thermal behavior, i.e. a steady-state solution with an intermediate value of temperature (Fig. 5) [27].

3.3 The Pressure Drop — Flow Rate Dependence

Let us now discuss the investigations of the flow pattern of the polymerization process in plug reactors. The operation of plug reactors, basically tubular, is usually considered in flows of polymerizing masses at $\eta(t \to \infty) < \infty$. The fact that an inhomogeneous distribution of rheological properties of the material over the volume arises necessarily in any real hydrodynamic situation is of fundamental importance for the analysis of the problem. As will be seen below, the distribution of the reactant concentrations and the viscosity over the section of the tube plays the main role for the tubular polymerization reactors. It is precisely these inhomogeneities which quantitatively affect the process pattern. Therefore, a correct description of the flow pattern must take into account the distribution of all characteristics of the reactive mass over the volume of the device as well as over its section. Any simpler model ("suspended") cannot in principle describe the main qualitative mechanisms of the flow or say anything about quantitative aspects. Not only the simplest models of stirred tank reactors and plug-flow reactors already considered in this paper, but also their modifications and combinations (dispersion model, cascade model, etc. [17]), are very restricted in their application for description of the polymerization process just because they "digress" from the description of liquid flows, i.e. they ignore deliberately the rheokinetics of the process. This is the main restriction and drawback of the model based on a combination of stirred tank reactors with drag flow [28], two-dimensional cascade [29] and similar models. In these cases, the accepted models and real technological situations are, essentially, adjusted in a semiempirical way by varying empirical constants in mathematical equations.

It is a usual practice to use averaged models (unidimensional where averaging is over the section, and β, η, T vary along the length of the tube, and zero-dimensional [suspended] where averaging is over the whole volume) in order to understand certain mechanisms of liquid flows including polymerizing liquids. Such an analysis is the first step in studying thermal mechanisms of the process, it gives an idea of temperature and concentration variation effects in the reactor and serves as a basis for further analysis. Finally, it is particularly important that on the basis of this simple statement we can investigate in detail such an important characteristic of the process as the depend-

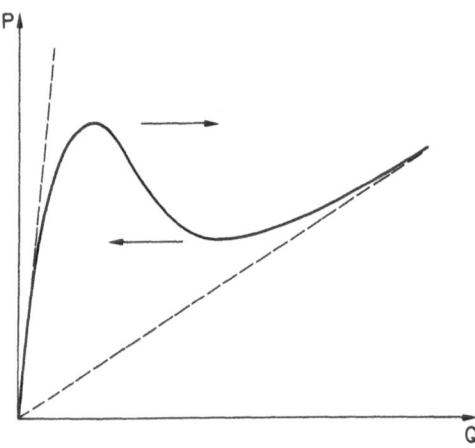

Fig. 6. Non-monotonic dependence of pressure drop P on flow rate Q.

ence of the pressure drop P on the flow rate Q (pressure drop-flow rate characteristic). Let us dwell on the function P(Q) in more detail.

If a liquid with a constant viscosity flows in a tube, the P(Q) function is described by Poiseuille's law: $P \sim \eta Q$. If the viscosity varies during the flow for some reasons (decreases with temperature, increases in the course of a chemical reaction), the linear Poiseulle dependence is violated and non-monotony of the pressure drop-flow rate curve may even appear (Fig. 6). Such non-monotony was first discovered while analyzing the flow of a hot inert liquid, whose viscosity strongly depended on temperature $\eta(T)$, in a cold tube [30]. As the flow rate increased, the colder, i.e. more viscous, liquid was carried out from the tube and the hotter, less viscous, liquid occupied its place. As a result, the integral (by volume) viscosity decreased and this may have caused a reduction in the pressure drop. The authors of Ref. [31] also discovered a reduction in P with the growth of flow rate during liquid flow with the strong $\eta(T)$ dependence. However, the reason was completely different; it happened due to an increase of the effective dissipative temperature as a result of viscous friction. Theoretical results of Ref. [31] were confirmed by experimental investigation of dissipative effects using a rotational viscosimeter [32]. The authors analyzed experimentally the dependence between the angular velocity (kinematic characteristic — analog of flow rate) and the applied torque (dynamic characteristic — analog of pressure drop for the flow in a tube). The authors of Ref. [32] not only found hysteresis effects similar to those predicted earlier in Ref. [31], but also proved this direct analogy and the possibility to transfer qualitative results obtained in a purely shear (Couette) flow to enforced flow in a tube. In all these works, it was assumed that if P = const is maintained, the states on the branch P(Q) with negative slope are unstable, which was confirmed by an analysis in Ref. [33].

Obviously, a sharp viscosity growth during polymerization may also lead to a reduction of P(Q) due to variation of integral viscosity with the flow rate

$$P \sim Q \int_0^L \eta(z) \, dz.$$

With the growth of flow rate from low values, the elements of the reactive mass with the longest residence time are carried out from the reactor, i.e. those that have reacted to the maximal depth and have the maximal viscosity. Instead, the reactor is refilled with a fresh mixture of minimal viscosity. The effect of non-monotony of the $P(Q)$ curve in isothermal flow of polymerizing liquid was independently discovered in Refs. [34, 35], and moreover, in Ref. [34] its qualitative experimental confirmation (instabilities during the operation of the corresponding multipass heat exchangers) was reported.

If the results of Refs. [30, 31] are of theoretical interest, the non-monotony of pressure drop — flow rate characteristic of a tubular reactor is essential for the organization of its operation, control, etc. Let us return to Fig. 6. The pressure drop is usually determined by hydrodynamic parameters of the tube. Let us consider what will occur in the system, if we gradually increase, this parameter. When the peak of the $P(Q)$ curve is attained, an abrupt transition occurs in the system (arrow in Fig. 6) from the steady state on the left-hand branch (with low flow rate and high conversion) to the state on the right-hand branch (large Q, small β). With an opposite change of the pressure, the transition from the right-hand to the left-hand branch will occur in some other place, i.e. hysteresis will be operative. In Ref. [33], the details of hysteresis and non-hysteresis transitions are studied. It is seen that there exists a qualitative difference between them (Fig. 7). The transition between the states on one branch is of a relaxation character (Fig. 7a),

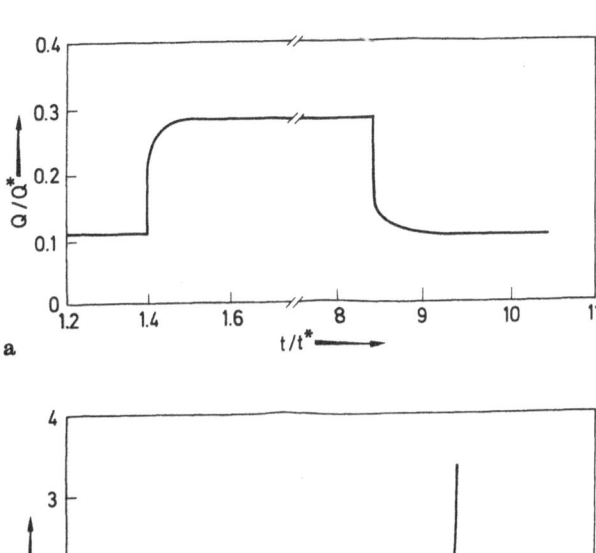

Fig. 7a, b. Unsteady transitions in the system upon a variation of the pressure drop (variation of flow rate with time): **(a)** relaxation transition (initial and final states are on one branch of the solution); **(b)** wave transition (from one branch of the solution to another)

while the hysteresis transition is of a wave character, i.e. the major part of the transition occurs with a constant velocity (Fig. 7b). Let us discuss one more result of Ref. [33] which is important in practice. If the process has broken away (for example, as a result of fluctuations of P) from the peak of the P(Q) curve, then by a repeated change of pressure drop during the transition the system can be returned to the initial position rather easily (for a period of time over a large interval of flow rate variation). If the break away took place from the minimum of the P(Q) curve, it was practically impossible to stop the starting transition.

In a number of works on the analysis of the P(Q) dependence for polymerization plug reactors [24, 26, 27], it was shown that all indicated reasons for the appearance of non-monotony of the P(Q) dependence may act jointly or independently, resulting in composite and complicated pressure drop — flow rate curves. Even the appearance of one more pair of extrema on the P(Q) curve (Fig. 8) complicates significantly the prob-

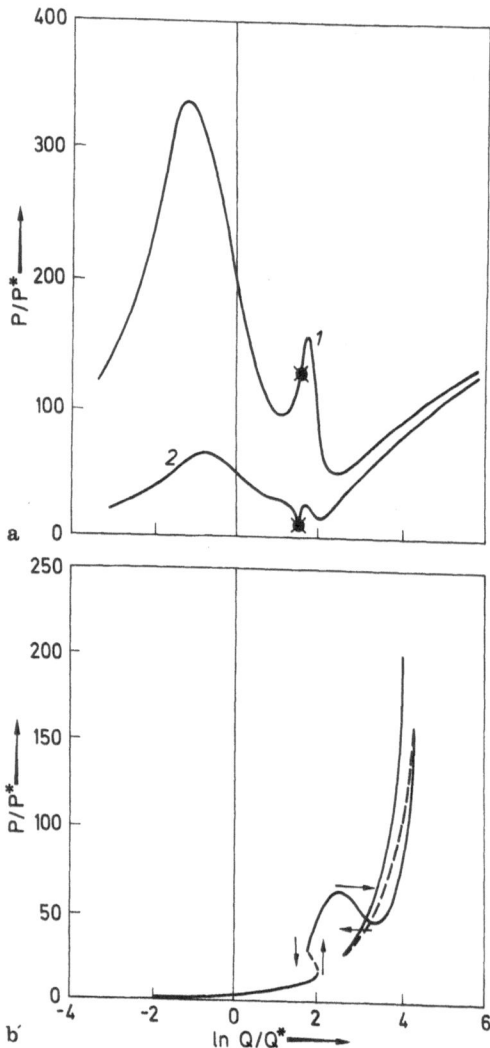

Fig. 8a, b. Types of dependences of P(Q) (dimensionless quantities): (a) the presence of a strong dissipative heat output, crosses denote the transition to a high-temperature process; _1_ — a more strong dependence η(β) than η(T); _2_ — opposite case; (b) high intensity of axial thermal conductivity; _dashed lines_ denote absolute unsteady states. _Arrows_ show possible hysteresis transitions upon variation of P(a) and Q(b)

lems on appearance and stability of steady states, transition relation between the branches number of steady states, etc. The authors of Ref. [24] have found one more reason for the appearance of non-monotony — the transition from one temperature behavior to another with the variation of the flow rate. It accompanies by a sharp rearrangement of temperature concentrations and viscosity profiles, and, consequently, the value of integral viscosity changes markedly. However, in principle, a competition is possible between temperature and concentration dependences of viscosity; for example, during polymerization in a solution. A detailed analysis of all possible versions of variations of the type and shape of the P(Q) curve has been carried out in Ref. [24], where the most complete statement of the problem for a unidimensional case has been discussed, and a significant role of heat conduction in axial direction (along the metal of the tube) has been taken into account.

In this case, unstable stationary conditions are possible. The variation pattern for P(Q) is very complicated and is determined by the $\eta(\beta, T)$ dependence, thermal factors (chemical and mechanical sources of heat, conductive and convective heat transfer and heat exchange through the wall of a reactor) and by the ratio of the rates in different chemical stages. Figure 8a, b shows curves P(Q) of a complicated form which illustrate this behavior. A similar analysis applied to a screw polymerization reactor has been carried out in Refs. [27, 36].

A few words on the stability of steady states of polymerization. This question arises immediately as soon as the multiplicity of steady-state conditions appears. It is well known that three solutions are possible in the flow of reactants. The general theory of thermal instability of reactors has been developed in detail in Refs. [16–20, 30, 31], and the theory of kinetic instability caused by peculiarities of the kinetic scheme (self-acceleration, gel-effect, etc. in Refs. [37–40]). The instability of steady states of polymerization plug reactors of a hydrodynamic nature is more interesting for the present paper. It can be assumed that the state corresponding to the negative slopes of the P(Q) curve are unstable if P = const is maintained [30, 33, 34]. At Q = const, all states are stable and realizable. The analysis of this problem in zero-dimensional formulation [41], for a reactor determined by only one value of T, β, η and a complex variable hydrodynamic resistance has shown that the slope of the curve is not an exhaustive stability criterion.

A strict mathematical analysis of this situation [42] confirms observations on the unstable operation of multipipe heat exchangers in which polymerizing mass flows [34]. Random small variations of flow rate are increased by an overshoot of a less viscous liquid; the residence time is reduced, the viscosity drops, while the tendency to an overshoot grows and, as a result, the system enters unstable operating conditions (the main flow goes through one or several tubes, while most of the tubes are plugged by slowly moving products).

It should be noted that all investigations of flow stability of polymerizing liquids are few in number and have been carried out up till now only for unidimensional problems. The problem of stability of steady rheokinetic two-dimensional flows to local hydrodynamic perturbations has not been discussed in the literature yet. Obviously the problem can be solved (the solution is difficult from the technical point of view), for example, by numerical methods solving the problem on unsteady development of the flow of polymerizing mass directly after a forced local change of the profile of the flow velocity.

3.4 Radial Distributions in Tubular Polymerization Reactors. Low-Temperature Process

A complete analytical examination of the role of distribution of the flow velocity over the radius of a tube is obviously impossible. A formulated problem for a complete description of the flow of rheokinetic liquid seems to be quite difficult and it is clear that the first steps in investigating a two-dimensional flow were based on very simple assumptions. In a number of works [43, 44], the authors took a fixed parabolic profile which is incorrect in principle for the flow of polymerizing media and leads to important mistakes. This is demonstrated very well in Ref. [45] where the possibility for styrene polymerization in a tubular reactor has been estimated: it has been shown that, if a real distribution of flow velocities and residence times over the radius is taken into account, the answer must be negative, in Ref. [44] however, a positive answer is obtained for an a priori parabolic profile.

One more approach [46, 47] describes the solution by giving a predetermimed profile of the flow velocity derived from experiments on heat exchange with the surrounding medium (the value of the Biot Number $Bi = \dfrac{\alpha r_0}{\lambda}$; r_0 is the radius of the tube). Such a model taking into account the thermal effects on velocity profiles is quite permissible for "conventional" reactions of low-molecular compounds [48], but cannot be applied directly to flows for which the viscosity growth is determined by the velocity of chemical conversion.

A number of interesting results were obtained on the analysis of two-dimensional flows of liquids whose viscosity depends only on temperature. An attempt to find out a heat wave of dissipative origin in Ref. [49] during the flow of an inert liquid has led to an obvious result: the initial perturbation ("wave") becomes attenuated as it propagates. The authors of Ref. [49] wanted to expand the analogies between the combustion processes and nonisothermal flows (earlier such analogies included "hydrodynamic thermal explosion", „hydrodynamic ignition and extinction", "degeneracy of hydrodynamic ignition", etc. [1]). However, an analog of the wave propagation of the combustion process was discovered for the isothermal hysteresis transition with the variation of pressure drop (Fig. 6 [33]).

Another similar curious effect is "hydrodynamic ignition" [50]. A larger velocity gradient near the walls of the tube yields a greater intensity of mechanical heat output, which results in an acceleration of chemical reaction, growth of temperature near a wall ("ignition"), which then propagates to the axis of the reactor. Evidently, this effect discovered theoretically may take place in practice during the flow of high-viscosity reacting liquid with a weak dependence $\eta(\beta)$. However, the authors of Ref. [50] try to transfer the obtained results to the flow of ethylene in a tubular reactor which, in our opinion, is highly questionable. For the turbulent flow of a monomer, the wall zone is not separated in a hydrodynamic sense, while for a laminar flow of a mixture of reactants the viscosity growth leads to a directly opposite tendency — a slowing-down of the flow near the walls.

In Ref. [51] a two-dimensional analysis was carried out and inclusion of the $\eta(\beta)$ dependence was made. However, the authors have not discovered any real effect of the viscosity growth on the mechanisms of the flow in contrast to all other works ot a similar nature. This discrepancy is apparently connected with the very weak relation between the viscosity and conversion assumed by the authors.

The investigation by Lynn and Huff [9] was the first one in which the true profile of flow velocities during polymerization in a tubular reactor (on the example of butadiene) was determined. The idea of the dependence of hydrodynamic fields on the viscosity growth during polymerization was clearly implemented. Starting from this model, the authors obtained sharply protruded profiles of flow velocities when the velocity on the axis of the tube exceed an average velocity by an order of magnitude. In another such model the following situation was simulated: the viscosity of the medium changes abruptly from the initial to the final case. It became possible to reduce the solution to one integral equation with self-similarity solution [52]. A complete statement and model analysis of the numerical solution were carried out in Ref. [53]. Different sides and versions of the solution of the problem were investigated numerically in papers [45, 54–58].

The most important result of these calculations is the following: If a laminar flow occurs, the condition for adhesion on the wall is fulfilled and the viscosity of the reacting liquid grows during the reaction at least by two orders of magnitude, then a strong deformation of the profile of the flow velocities occurs in the reactor and concave profiles of the flow velocity appear, which are completely anomalous from the point of view of conventional hydrodynamics. The reason for their appearance is the action of a positive feedback. A small flow velocity near the wall leads to a long residence time of these elements of the liquid in the reactor, i.e. to higher conversions and a strong viscosity growth of wall layers and, consequently, to slowing-down of their flow. In the axial zone that same positive feedback has the opposite effect. Originally,

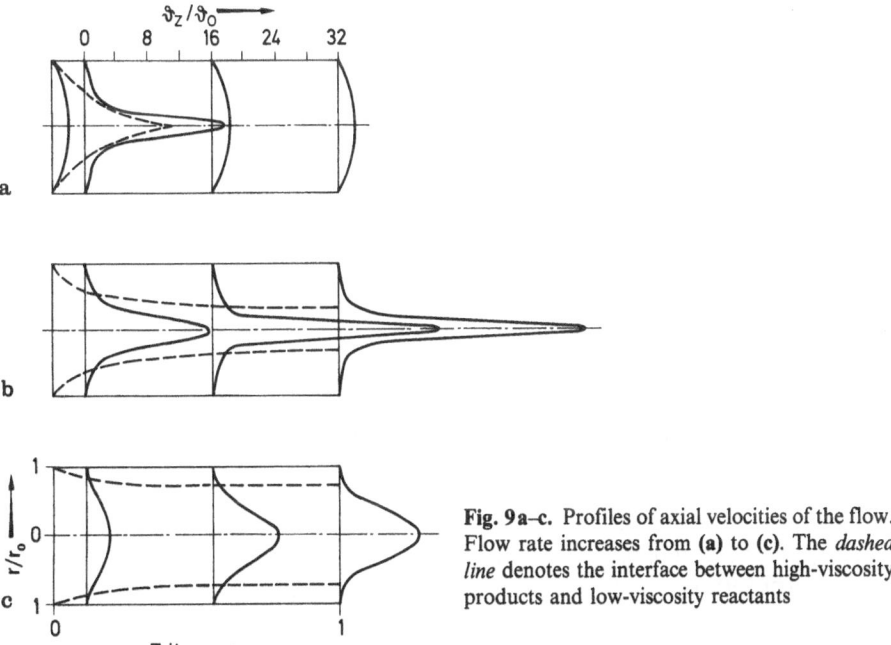

Fig. 9a–c. Profiles of axial velocities of the flow. Flow rate increases from **(a)** to **(c)**. The *dashed line* denotes the interface between high-viscosity products and low-viscosity reactants

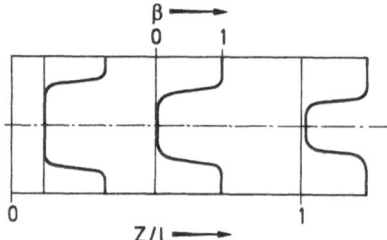

Fig. 10. Profiles of monomer conversion in the reactor

the liquid on the axis moves faster and has a smaller residence time and both conversion and viscosity are lower as compared to wall layers. In the final result, in addition to characteristic profiles of flow velocities (Fig. 9), two rather clearly and sharply demarcated zones of flow are formed: a zone of slowly moving products near the reactor's wall and an axial zone of a fast stream of reactants (Fig. 10), which is in agreement with the model approach of Ref. [52]. With small flow rates, the stream of reactants resides completely within the reactor; when the flow rate increases, it extends and forces through the zone of the products. Further increase in the flow rate leads to the expansion and smearing out the stream zone and the reduction of the layer of products at the reactor walls (Figs. 9, 10).

Thus, the most typical feature of the hydrodynamic pattern of polymerization in a tubular reactor is an overshoot of reactants in the axial zone. The greater the viscosity growth during reaction, the more the profiles of the flow velocities extend. When the ratio between the final and initial viscosities $\eta_f/\eta_0 = 1000 \div 3000$, the velocity at the axis may exceed the average velocity by a factor of $20 \div 40$.

We note that in ordinary hydrodynamic problems, where the viscosity of a liquid does not vary with time due to chemical reaction, the profiles of the flow velocities described above (i.e. strongly extended with a point of inflection differing sharply from the usual Poiseuille parabola) are unstable. If we consider such a profile as initial in a tube with a liquid with constant viscosity (or, for instance, a non-Newtonian liquid), then in the course of the unsteady process the profile of the flow velocities will necessarily approach a steady parabola (or another convex curve for a non-Newtonian liquid). In the case of a rheokinetic problem the profiles with an inflection point evidently lead to a quite stable steady-state solution. The stipulation "evidently" is made because the stability of such flows towards two-dimensional perturbations has not been analyzed yet, but indirect arguments and experiments support the conclusion stated above. It is useful to note that the greater the ratio η_f/η_0, the more difficult is the numerical solution, and the more detailed and small should be the partition of the reactor space. The most convenient and economic is a mobile net which "arranges itself" to a variable hydrodynamic pattern of the process. The steeper is the velocity gradient (or the gradient of the conversion) over the section of the tube the greater must be the number of nodes. Even the simplest constant network 20×40 fully ensures the accuracy of calculations with respect to residence time (and to conversion) with 10% relative error at $\eta_f/\eta_0 = 100 \div 400$ [54, 57, 58].

A sharp distortion of profiles and overshoot of reactants along the axis are negative features of the operation of a tubular polymerization reactor. Its utilization is more or less effective only if the viscosity does not grow too strongly. If this growth amounts to at least two orders of magnitude, no external forces (change of temperature, organization of recycling, separate supply of reactants, etc.) will help to get rid of the phenomena described above: non-uniformity of conversion, extension of profiles, overshoot of reactants. Obviously, to attain a given and sufficiently high conversion, a considerably lower flow rate is required or, at a fixed flow rate, a considerably longer tube (than is the value obtained from the calculation with respect to an average residence time $L/(Q/\pi r_0^2)$). The results of Ref. [54] show that for such flows one can write a relationship of the type of Poiseuille equations for a small role of thermal effects (low-temperature polymerization conditions):

$$L^* = \gamma_1 L_{min}^* = \gamma_1 \frac{Q}{\pi r_0^2} t^*$$

$$P = \gamma_2 \eta_0 \frac{8L^*Q}{\pi r_0^4} = \gamma_3 \eta_0 \frac{8Q^2 t^*}{\pi r_0^6}$$

(L_{min}^* and t^* are the minimal length of the tube and the residence time calculated via the plug-flow model; L^* is the actual length of the tube (taking into account the deformation of the profile of flow velocities) which provides the required conversion corresponding to the time t^*).

Here, $\gamma_1, \gamma_2, \gamma_3$ are the calculated coefficients determined by the ratio between the initial and final viscosities. The value $\gamma_1 > 2$ shows just how much faster the liquid flows in the axis direction than is the average value. The coefficient γ_2 represents the viscosity growth during the reaction and is close to the ratio between the final and initial viscosities. The values of $\gamma_3 = \gamma_2\gamma_1$ and of γ_3 are usually very large. For example, if the viscosity increases 125 times, $\gamma_3 \approx 600$ [54]. Using these relationships, it is possible to transfer the results obtained for small flow rates to the flows with large values Q and there is no need to carry out long and complicated calculations. This simple procedure makes modelling and optimization of the operation of a tubular reactor much easier.

The analysis of the effect the form of the $\eta(\beta)$ dependence exerts on the flow pattern [59] has confirmed that for the case of strong viscosity growth, just the ratio η_f/η_0 is the main parameter of the problem. The way by which the viscosity changes especially in the region of high conversions does not essentially affect the development steady states of the process.

The fact that the results obtained earlier and described above were experimentally confirmed was of great importance. The literature does not contain experimental investigations of rheokinetic problems in which the distribution of flow velocities or residence times (conversion) at the output of the tubular reactor would be studied. Therefore, the results of the experimental investigation of hydrolytic polymerization of dodecalactam in a pilot tubular reactor and the comparison of these results with computation [58] deserve a more detailed presentation.

The viscosity of the reactants varied by a factor of $200 \div 400$ in the course of the process, which led to a sufficiently strong distortion of the profile of flow velocities and deviation of all quantites from the values calculated on the assumption of plug flow or fixed parabolic $V_z(r)$. The experimental data and calculated values of average con-

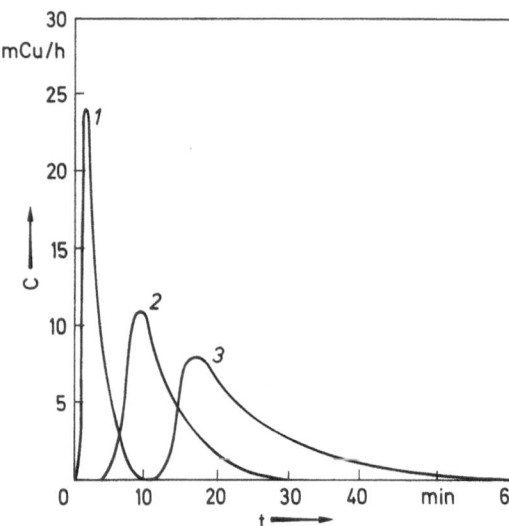

Fig. 11. Response curves: *1, 2* — experimental curves at the input and output of the reactor; *3* — calculation for the flow with a fixed parabolic profile of flow velocities, output of the reactor

version β and minimum residence time t_{min} were compared. These values were calculated using the following formulas:

$$\bar{\beta} = \frac{1}{Q} \int_0^{r_0} \beta(r) \, 2\pi r V_z(r) \, dr \, .$$

$$t_{min} = \int_0^L \frac{dz}{V_z(r = 0)} \, .$$

Calculated and experimental β and t_{min} values were compared for different rates, initial temperatures and concentrations. The maximum deviation in the average conversion did not exceed 15%. The values of t_{min} were measured by the radio tracer method and the deviation did not exceed 10%. The authors also used the models of plug and Poiseuille flows and showed that the application of these models to rheokinetic flows is incorrect in principle: divergences in values of the average conversion amounted to 50% and those in the minimum residence time exceeded 100%. The conclusion on the incorrectness of idealized models of flow is also confirmed by the response curves for the experiment with a radiotracer (Fig. 11): quantitative and qualitative divergence of response curves is obvious in the experiment and in calculations according to the model of Poiseuille flow with constant viscosity (cf. curves 2 and 3).

3.5 Peculiarities of "Hydrodynamic" MMD

Recently, theoretical investigations were carried out of the effect of hydrodynamics of polymerizing liquids on molecular properties of the product — its molecular-mass distribution (MMD) [60–63]. The authors solved the rheokinetic problem, formulated at the beginning of this review supplementing it with the equations for the MMD moments. The distribution function $r_w(j)$ was determined via the equation

$$r_w(j) = \frac{1}{Q} \int_0^{r_0} q_w(j, r) \, 2\pi r V_z(r) \, dr \, , \tag{7}$$

where $q_w(j, r)$ is the weight-fraction distribution function in each microvolume determined from purely kinetic considerations. Examining the case of "living" anionic polymerization for which MMD is described by the number-fraction Poisson function

$$q_n(j) = \frac{\tau^{j-1}}{(j-1)!} e^{-\tau}$$

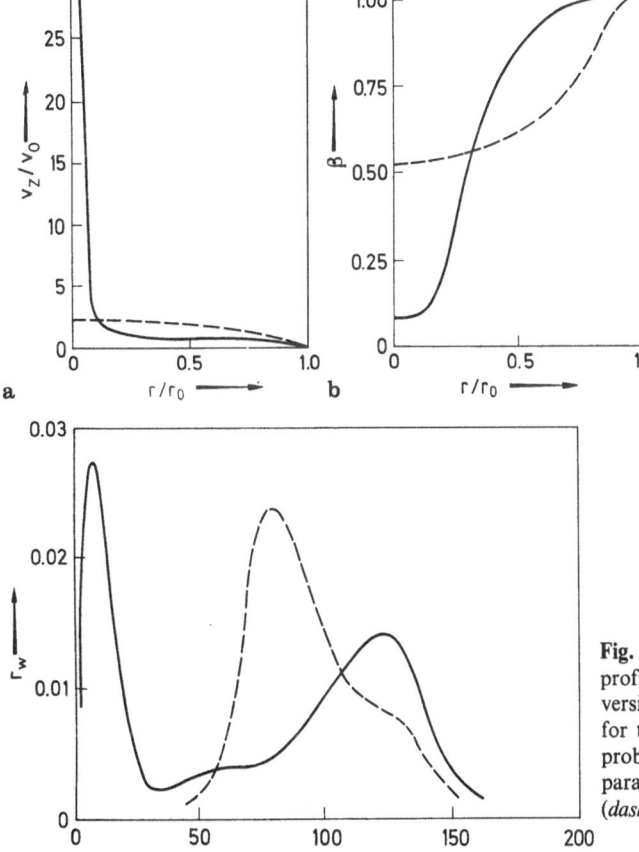

Fig. 12 a–c. A comparison of the profile of flow velocities (a), conversion profile (b), and *MMD* (c) for the solution of a rheokinetic problem (*solid lines*) and a fixed parabolic profile òf flow velocity (*dashed lines*)

where τ is the reduced time, simple expressions for the MMD moments λ_0, λ_1,, λ_2 were determined:

$$\lambda_0 = \frac{1}{1 + \tau} \; ; \quad \lambda_1 = 1 \; ; \quad \lambda_2 = \frac{1 + \tau^2 + 3\tau}{1 + \tau} .$$

Using these relations and calculating the profile $V_z(r)$ by Eq. (7) one can determine the properties of MMD and the MMD itself at the output of the tubular reactor.

The presence of two zones of flow, which was discussed in detail above, being about the formation in the output of the reactor of two fractions of a polymer with a very different average residence time for each fraction. In case of a "living" polymerization when an average degree of polymerization grows proportionally to the reaction time, the superposition of two such fractions with clearly different molecular weights leads to a transformation of the Poisson MMD, initially very narrow, into a significantly different and even bimodal distribution. Figure 12 confirms the fundamental difference of the solution of a rheokinetic problem for Poiseuille flow.

Experimental data concerning the effect of hydrodynamics of the flow in a tube (Reynolds Number) on the MMD of a polyurethane have shown [64] that the lower the flow velocity, the wider is the MMD and the greater is the average molecular mass of polyurethane. Obviously, in the case, the form of MMD is determined by the effect of the curing reaction.

3.6 Conclusions

We can now sum up the outline of low-temperature conditions for the flow of rheo-kinetic liquids during polymerization in tubular reactors. It seems to be quite clear that in the physical pattern of the process the viscosity growth plays a principal role. The system of equations is formulated which describe the process correctly and adequately [Eqs. (1)–(5) of this review]. Numerical methods are suggested for solving the hydrodynamic problem of flow of a polymerizing liquid in a tubular reactor. The methods have been developed independently by a number of groups. Ignoring the effect of the viscosity growth on the mechanisms of flow of rheokinetic liquids is, in principle, incorrect. Theoretical predictions have been confirmed experimentally. Empirical equations (6) are formulated; they make the calculation and optimization of a reactor's operation easier. A general approach to finding "hydrodynamic" molecular mass distribution is formulated and certain specific solutions are obtained for particular kinetic schemes. Thus, we can speak about a complete statement of the problem of flow of a rheokinetic liquid and about its solution.

We should point out the low efficiency of single-flow tubular reactors: only a small portion of the reactor volume is used effectively, an essentially inhomogeneous flow structure is formed, the product moves slowly through the reactor, etc. Moreover, technological difficulties sharply increase with increasing of the final and initial viscosites and at $\eta_f/\eta_0 > 10^3$ the use of tubular reactors without additional technical improvements is inexpedient. Forced agitation with the help of static mixers, a sequence of tubular reactors with agitated reactors, etc., may be among such artificial improvements. In any case, the efficiency of these or other devices can be estimated by the rheokinetic method.

4 "Front" Polymerization

Polymerization processes have, as a rule, sufficiently high exothermicity and therefore they may occur with self-heating under "quasi-adiabatic" or high-temperature conditions. The mechanisms of nonisothermal polymerization are considered in a number of papers and monographs (cf. e.g. the review [65]). Comparatively recently, it was discovered that besides volume self-heating over the whole mass of reactants the process may develop layer-by-layer: due to heat conduction the heat from the hot zone with intensive reaction is transferred to the neighboring layer of reactants, sharply accelerates the process in this layer which heats up quickly, and so on. The authors of experimental work [66] who found out this process called it "front" polymerization [66] because it is analogous to the combustion processes where the front (combustion wave) propagates by the same mechanism. Later, front polymerization was investigated in detail experimentally and theoretically with the help of the mathematical approach of the combustion theory [19], for example, in Refs. [67–69].

The analysis of high-temperature polymerization in a tubular reactor including the distributions of temperatures, concentrations and flow velocities over the reactor radius confirms the conclusion about the front propagation of this process. This kind of problem was first theoretically studied in Ref. [70] in which a comparison with the data of an independent experiment [71] was made. The thermal pattern of the development of the process [71] was analyzed in detail. The situation where the fast forming polymerization front is perpendicular to the flow direction was analyzed and it was shown that a sharp growth of viscosity makes the process unsteady with a transition to a low-temperature condition. Let us consider the development of such a process in detail.

The transition of the process into a high-temperature regime is possible only if flow rate is not too large and not too small and the value of the Frank-Kamenetskii Number δ is greater than the critical value

$$\delta = \frac{r_0^2 q E}{\lambda R T_0^2} K(T_0) > \delta_{cr}$$

where $K(T_0)$ is the reaction constant at temperature T_0 of the medium surrounding the reactor. The Frank-Kamentskii Number is a complete analogue of the Semyonov Number $\varkappa$ for distributed systems (the ratio between the intensity of heat output and the intensity of heat removal).

When the process starts at the initial temperature T_0, a flash occurs after a certain period of induction in the "tail" part of the reactor. The reactants are heated up to high temperature (by $100 \div 150\,K$ for a typical polymerization process). Later, a polymerization front is formed quickly and the greater part of the chemical transformation takes place in the region with the maximum temperature. The development of the process is shown in Fig. 13.

The next stage of the process is the motion of the polymerization front in the opposite direction to the flow of reactants in order to reach the state when the feed rate of the reactants V_f is comparable with the "combustion" rate U_c [19] (reaction rate at the front):

$$V_f \approx \frac{Q}{\pi r_0^2} \approx U_c = \frac{C}{\varrho} \frac{RT_c^2}{E} \sqrt{2aK(T_c)} \,.$$

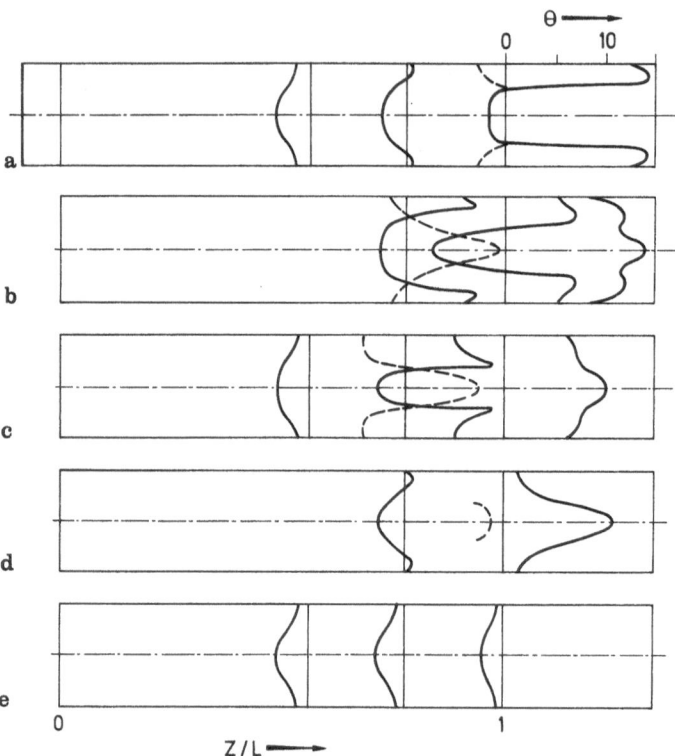

Fig. 13a–e. Development (from **a** to **e**) of unsteady high-temperature polymerization (*solid lines* denote temperature profile, *dashed lines* — the position of polymerization front)

Here, a is the thermal diffusivity of the reactive mass, T_c is the combustion temperature (the maximum temperature of the front), $K(T_c)$ is the reaction constant at $T = T_c$. If this condition is met, the process becomes stabilized. In the experiment [71], the stabilization stage (constant position of the front) amounted to tens of minutes. All these stages are similar to the corresponding development of the combustion process [19] and if the viscosity of a substance were not increasing, the development of the process would be completed at this stage. It is obvious that the analysis of the polymerization front propagating in a stationary medium [66–69] does not yield any qualitatively new results as compared to combustion.

However, rheokinetic effects cause the development before the front with exactly the same pattern as under low-temperature conditions (Figs. 9, 10). The products accumulate on the walls and in the axial zone the flow is accelerated, i.e. the feed rate of the reactants increases. As a result, the $V_f = U_c$ equilibrium is violated, the front line is distorted and its central part is displaced towards the output. Consequently, the temperature becomes lower, the rate of "combustion" drops, and the feed — combustion equilibrium is violated still more. Also, the front region is cooled down and is transferred out of the tube. Therefore, for a rheokinetic liquid (polymerizing medium with a sharp viscosity growth), a low-temperature condition for the process is the only steady-state solution. The polymerization front normal to the flow can exist only as an unsteady state and this solution is unstable.

A search for steady-state solutions for front polymerization based on the combustion theory has shown [72] that in principle such steady-state conditions exist. However, the front does not have the form of a plane, but of a strongly extended conical surface of the same form as a products — reactants boundary under the low-temperature condition (Fig. 9, 10). Hence, the whole positive technological effect of high-temperature process (front polymerization) vanishes because one obtain neither a reduction of the reactor length nor a necessary average flow rate, nor an improvement in utilizing the volume of the device. Because of these negative results of theoretical calculations for a tubular reactor, a group of investigators suggested continuous-flow reactors for front polymerization with new forms — spherical and cylindrical — with radial feed of the reactants [73]. Later, thermal calculation was carried out for such reactors but the rheokinetics properly was not included. Some interesting ideas about the possible technological application of front polymerization (the production of new composite materials, etc.) are given in Ref. [74].

Another variant of intensive polymerization processes consists of agitating the reactants in the zone of high temperature increase. Such agitation due to natural convection is possible in a tubular reactor made in the form of a horizontal coil [75]. The use of one-screw [36, 76–78] or double-screw [79–82] extruders as reactors is very interesting since these devices make it possible to provide (as compared to the usual tubular reactors) a far greater homogeneity of products at the output, a more intensive heat removal, the possibility of stabilizing the "hot zone" of the process (due to circulating cross flows in a channel) as well as the possibility of eliminating the intermediate stages of the production process "raw material — finished product". The drawbacks of screw polymerization devices are a comparatively low productivity and constructive complexity which increases specific capital outlay.

A two-dimensional analysis of the flow in a screw channel begun from investigating the Couette flow of polymerizing liquid between the rotating cylinders [76] and the flow mechanisms in a screw channel [77, 78] were analyzed. It was shown that the viscosity growth during polymerization (dependence on molecular mass) significantly distorts the profiles of temperature, flow velocities and, especially, conversions over

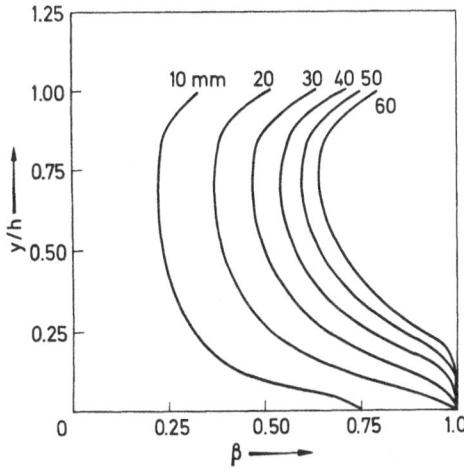

Fig. 14. Conversion profiles in a screw reactor at different distances from the output of the reactor (figures near curves denote the distance from the output)

the width of a channel as compared to the case of flow of a liquid with a constant viscosity or of an inert liquid whose viscosity varies with temperature. The same phenomena are observed as in a tubular reactor, i.e. the flow velocity profiles have an inflection point and nonuniform distribution of conversion ("trough-shaped" profiles
of the conversion). Nevertheless, the nonuniformity of conversion profiles in a screw
reactor is essentially lower than in a tube (cf. Figs. 10, 14); moreover, it can be controlled both by varying the rotational velocity of the screw and by varying the temperatures of the working surfaces of a cylinder and a screw [78].

The studies of the polymerization process in a double-screw extruder have just
started. A complete mathematical statement of the problem has not been formulated
yet, the process is being studied experimentally [79–82], sometimes empirical models are used. In Ref. [81], a series of continuous stirred tank reactors (CSTR) is
suggested as a model or experimental residence times are used to construct a model
of a plug reactor. A very important characteristic of the process determining the properties of the final product (MMD) was measured — the distribution of residence
times of the reactive mass in the extruder. These data were related to the technological
parameter of the process (temperature, rotational velocity) but, unfortunately, were
not exposed to rheokinetic analysis. Nevertheless, the effect of the residence time
distribution on the MMD of the final product was confirmed by a direct experiment
[82].

5 Flow of Curing Liquid

Only recently have the investigations been started concerning the flow of curing
liquids, i.e. rheokinetic media losing the ability of viscous flow as a result of chemical
transformations. The investigations are directly connected with a new trend in polymerization engineering — the RIM-process. When moulds are filled with curing oligo-
meric compositions, a number of specific problems arises caused by the growth of
viscosity up to a complete loss of the ability to flow during gelation (network-formation
or solidification).

A step change of viscosity at critical conditions to infinity (3) is a convenient model
for investigating the flow of a curing liquid. The simplest model employs the most
important physical property of the process. In a number of papers [83–87], the dependence of the induction period t*, the time during which a reactive substance retains
the ability to flow, on the rate of shear $\dot{\gamma}$ is studied.

Nonisothermal flow of a curing model liquid in a flat slit is discussed in Refs.
[88, 89] as applied to the casting of rubber stock under the conditions of curing.

In all these cases, the condition of gelation (curing) for non-isothermal flow was
generalized with the help of a generalized criterion similar in structure to the well-
known criterion (according to Baily) of long-term strength under a variable load
operation [90]. At a temperature T_0 and induction period t_0^*, it is assumed that
under arbitrary temperature conditions $T(t)$ the value of the induction period t_T^*
is expressed as an upper integral limit in the formula

$$\int_0^{t_T^*} \exp\left\{\frac{E}{R}\left[\frac{1}{T_0} - \frac{1}{T(t)}\right]\right\} dt = t_0^*.$$

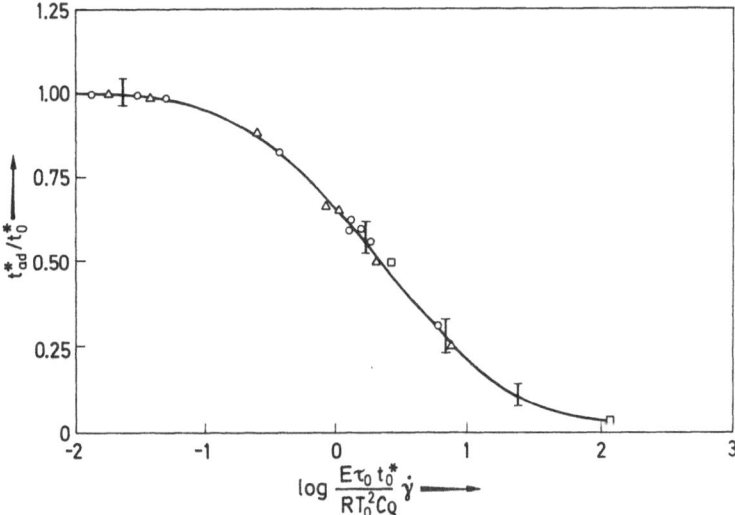

Fig. 15. Dependence of the reduced induction period $\tilde{t}_i^*$ on the dimensionless shear rate $\dot\gamma$

Solid line denotes the theoretical calculation using the formula $\tilde{t}_0^* = \dfrac{1}{\Gamma} \ln (1 + \Gamma)$. *Squares, Circles,*

triangles, denote the experiment for a silicone resin, for different T_0 423 K, 443 K, 463 K), bars denote a phenol-formaldehyde resin

Under the assumption of adiabatic conditions of flow, it was possible to obtain a simple analytic solution of the problem described by the formula

$$\tilde{t}_T^* = \frac{I}{\Gamma} \ln (1 + \Gamma).$$

where Γ is the dimensionless rate of shear, expressed as $\Gamma = E\dot\gamma\tau_0 t_0^*/RT_0^2/\varrho C$, where E is the activation energy, τ_0 is the initial shear stress; the value of $\tilde{t}^*$ is the reduced value of the induction period (assigned to $\tilde{t}_0^*$). The last formula shows that the theory predicts a sharp reduction of the value of the induction period with an increase in the shear rate observed experimentally (Fig. 15). Thus, the practically important effect of reducing the time of induction is determined by dissipative warming-up at high flow velocities.

Therefore, it is necessary: (1) to calculate real temperature fields in the production equipment during flow of curing mixtures with an allowance for specific rheological properties and real hydrodynamic situation; (2) the temperature fields in the techno-logical process are to be strictly controlled. These results stress the role and importance of mechanical sources of heat and the necessity to take them into account in all product-ion processes where high-viscosity polymers or polymerizing liquids flow.

Consideration of a two-dimensional flow pattern was a fundamental step in the analysis of the flow of rheokinetic curing liquids, as well as in the case of a sharp but limited viscosity growth. The loss of the possibility to flow leads to a stronger distortion of all characteristics over the section of the channel as compared to a usual pattern and to new qualitative regularities. This was demonstrated by Vaganov [91], where the analysis of different hydrodynamic, physical and chemical situations was successively performed using the approach developed earlier [52, 59]. This approach makes it possible to reduce the system of equations describing the flow of a liquid, when viscosity changes suddenly ($\eta \to \infty$), to one integral equation. The most important result of the theoretical analysis is that, if one assumes an ideal adhesion to the walls of the reactor (this is the most common boundary condition: $V_r(r_0) = V_z(r_0) = 0$), then a steady flow becomes impossible. This is connected with a continuous growth of a solidified (having lost the ability to flow) layer at the wall of the tube, which leads to a continuous contraction of the channel. A complete clogging of the channel takes place in a limited time, i.e. there is always a certain period of flow, but then a clogging, immobility emergency.

Unsteady development of such a flow was also investigated numerically and analytically in Ref. [92] both for a constant flow rate and constant pressure drop. In the analytical approach, it was considered that the flow stops after a ten-fold viscosity growth. A step model of viscosity variation was also used, but of more complex form than is assumed by Eq. (3). A comparison with the experiment has shown that the analytical model describes better longer periods of time, while the numerical approach "works" better in the initial stages. This indicates only an unsuccessful choice of the computational algorithm and demonstrates a typical process of error accumulation during calculations caused first of all by a too rough fixed two-dimensional network (17×13 in the most exact case).

A rapid advancement in the analysis of different types of flows as applied to the RIM-process has led to a new trend in chemical hydrodynamics or rheokinetics. Together with numerous works investigating specific flows of curing liquids, a number of generalizing investigations is being published. In Refs. [93, 94], the authors have analyzed the kinetics of curing, rheology, thermal and physical and other properties of a number of thermosetting resins. Different mathematical models of reactive casting and their applicability were discussed [95]. The general questions concerning the methods for analyzing rheokinetics and hydrodynamics of the process, peculiarities of heat transfer and constructional setting were formulated in Ref. [96].

The effect of polymerization kinetics and curing mechanism on the process of mould filling, pressure profiles at different conditions of viscosity variation, geometry of moulds, etc., is given in the review [97]. The review [98] is devoted to problems of interaction of heat and mass transfer during processing of polymerizing materials and to the generalization of earlier studies.

A characteristic feature of flows in forming items from curing resins is the presence of a free volume in the mould being filled, i.e. the presence of a freely moving surface of material. This feature together with the loss of ability to flow leads to a peculiar behavior of the medium at the front of the flow — the so-called fountain effect (Fig. 16). Its essence is rather simple and consists of the liquid from the central zone flowing much faster than the mass on the whole. A substance solidifying at the edges (walls) of the mould draws literally the layers from the central zone which turn solidify by the time they reach the walls. Because of viscous forces, they "draw" new portions of the liquid from the central zone of feed of reactants.

This phenomenon was analyzed for the first time in a quasi-one-dimensional approximation [99] on the assumption of a simple spreading of the liquid on the front by displacing it near the the walls of the channel, i.e. the two-dimensionality of the

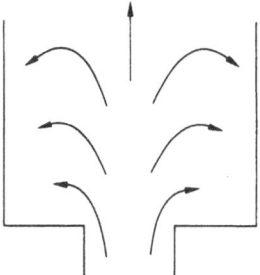

Fig. 16. Stream-lines at fountain effect

phenomenon was not investigated. However, this simplification did not interfere with the most important conclusion about the enormous influence of the fountain effect on the residence time distribution in the channel and, consequently, on the properties of the final product. The analysis of the flow profiles has shown that there is an analogy with the flow in polymerization tubular reactors, namely in a sharp deformation (extension) of the form of the profiles of longitudinal flow velocities. For example, in a calculation in Ref. [93], the maximal flow velocity exceeded the average velocity by a factor of 8.

The problem on the flow of a curing liquid in a flat slit was solved more strictly in two-dimensional problem [100]. The concept of the gel time, t_g, reached for a certain critical conversion — was used (an analogue of the induction period t^* [83–87]). Other authors solved the most different versions of flows and carried out specific calculations of the hydrodynamics of the process for unit production equipment and various moulds. From the point of view of the present paper, of some interest is the theoretical analysis of the flow in a rectangular channel [101], where the viscosity variation during the process was described by the following formula:

$$\eta = \eta_0 \exp \frac{U}{RT}\left(\frac{c_0 - c_g}{c - c_g}\right)$$

where c, c_0, c_g are respectively, the current, initial and gel point concentrations. It has been shown that a gradual transition occurs from the initial parabolic profile of the flow velocities to the concave one of the same type as that discussed in Sect. 3.4.

Flow in the nozzle of a moulding machine [102] and its pipeline [101], filling and curing in a casting mould [103], in a semi-circular tank [104], rotating mould [105], etc., was considered.

In a number of experimental works, e.g. [106–108]), the effect of the process hydrodynamics (Reynolds Number, Re) on mixing and quality of the obtained products was studied and it was concluded that the higher the value of Re, the better is the mixing. The equipment has been developed for determining the zones of bad mixing [107] and recommendations have been given on the construction of the equipment to eliminate poor mixing [108, 109]. Evidently, further progress in this field can be reached by formulating a mathematical model which would correlate the hydrodynamical pattern of the process with the quality of the resulting product. This problem is similar to modelling of the flow of a rheokinetic liquid in a stirred tank reactor.

An exact statement of the problem, as well as methods for its numerical solution,

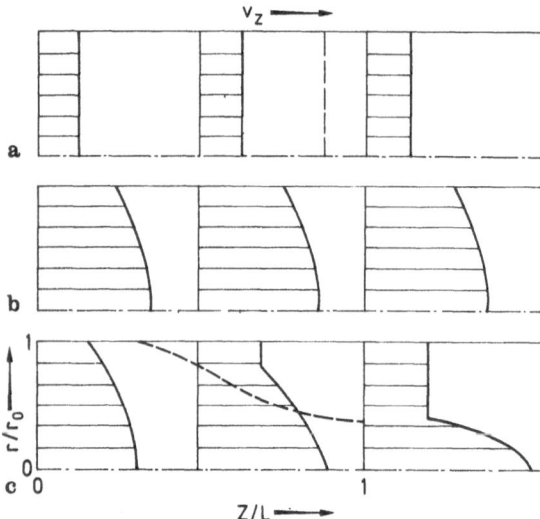

Fig. 17 a–c. Profiles of flow velocities of a curing liquid in the presence of a wall slip (**a, b, c** are the different conditions of flow, *dashed line* is the liquid-solid boundary)

and examples of calculation supported by experimental data make it possible to speak about creation of the theoretical basis of the RIM-process. The results generalized in Refs. [93–98] indicate that a new technology has been developed which successfully combines the process of production and processing of materials.

A further development of rheokinetics was started by the discussion of the flow of a curing liquid under the assumption that a solid material may slide along the wall of a channel [110, 111]. This is especially important since the steady state of the system becomes impossible. Theoretical investigations have proceeded from the simplest model assumptions: the viscosity dependence given by Eq. (3), i.e., the viscosity is constant during the induction period t* and at $t \geqq t^*$ the substance moves like a solid. It was also assumed that the slippage velocity was proportional to the shear stress at the wall. It has turned out that under the conditions of two-dimensional flow there are three possible versions of the motion of reactants shown in Fig. 17. We note that for the existence of steady-state solution the assumption is necessary that on the wall of the reactor there is always a solid phase which moves with a constant velocity.

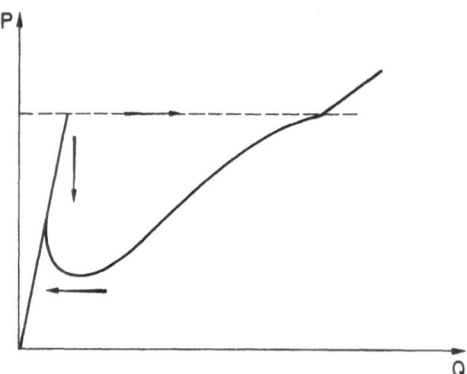

Fig. 18. The P(Q) dependence in the presence of a wall slip

The most curious result of this analysis was the P(Q) dependence for such a flow (Fig. 18) and other related results. The profiles of the flow velocities in Fig. 17a correspond to small rates — all substance has managed to become cured during the residence time in the tube — the left-hand segment of the line P(Q) in Fig. 18. Figure 17b shows the case of high rates, the liquid falls through the tube and has no time to react (except for a thin layer at the walls) — the right-hand segment of P(Q) in Fig. 18. Figure 17c shows a two-phase flow — the intermediate branch of the pressure-rate curve.

In a strictly isothermal flow process, three steadly-state solutions of the problem are possible which correspond to one value of the flow rate. This is the result of the joint influence of sliding and curing. From Fig. 18, it is seen that the transition from the flow of the type "all reacted completely in the tube" (Fig. 17a) to a two-phase flow (Fig. 17c) with the flow rate variation takes place abruptly accompanied by a complete rearrangement of the flow structure and of residence time distribution. If the pressure drop increases, such a transition becomes generally impossible, the flow passes into the type "nothing has managed to react" (Fig. 17b). In case of a decrease in flow rate or pressure drop, the process can reach the states shown in Fig. 17a only through an abrupt transition. The results obtained in the analysis agree at least qualitatively with the experiments.

6 Conclusion

The conclusion of a review is a "review of the review", i.e. an abstraction of the third order, if a theoretical analysis of a specific problem is considered as an abstraction of the first order and the review of such analysis, i.e. the present work on the whole, as an abstraction of the second order. Such an approach reminds one of a view of an artist working in the field of "nonrepresentational art", who escapes from the surrounding practical world. It is hardly probable that this would be fruitful in the present case. Therefore, concluding the discussion of the results of analyzing the flow of polymerizing liquids ("rheokinetic" liquids), we would like to make only a few general remarks.

In our opinion, the state of investigations in the field under consideration may be characterized by a phrase which is often used in any review "the basis of the theory has been created and the main trends of analysis have been developed". It is clear how to solve such problems but there exist complications: introduction of new stages of the process (for example, phase transition), complication of geometry, allowance for more complicated kinetic and rheological dependences etc. Basically, there are no obstacles to solving more and more complicated problems. There exist mathematical statements of the problems, methods of their solution, the physical pattern of the processes is clear, the most important physical factors and parameters of the process are revealed. Along with it, the theory has got ahead of the experiments. Experimental investigations of the flow of rheokinetic liquids can be counted on the fingers of one hand. Undoubtedly, a thorough experimental investigation is one of the urgent problems of rheokinetics. Besides, experimental data and results of investigations of many specific materials are lacking for the solution of specific problems and calculations of real technical situations. Thus, the shortage of experimental material is the most acute problem of the hydrodynamics of rheokinetic liquids.

We think that it is an urgent matter to solve concrete problems taking into account all possible peculiarities: different empirical relationships (rheological, kinetic), different geometry, various types of representing boundary conditions, organization of heat and mass transfer, etc. The absence of investigations of the effect of non-Newtonian properties of a liquid on the flow mechanisms of rheokinetic media seems to be a gap in the field of theoretical analysis.

The most important problem of rheokinetics remains the analysis of distribution of the residence times, especially for the stirred tank reactors. To develop a reasonable hydrodynamic model, which is physically proved and mathematically simple and takes into account the effect of a sharp viscosity growth is the main problem.

For the RIM-process it is necessary to create a model and formulate the problem which would enable us to correlate hydrodynamics of the process with uniformity of the resulting material and product, i.e. with the quality.

The important problems include the control of polymerization reactors containing rheokinetic liquids. These problems have not been solved in many respects even for more simple situations. Both mathematical modelling and physical understanding of the process are the key problems [112].

In terms of the techniques of mathematical analysis, the investigation of the stability of flow conditions towards multidimensional hydrodynamic perturbations is an important problem. It is clear that one of the ways to treat this problem is numerical analysis of similar unsteady-state conditions where a disturbed steady state is considered as the initial condition.

An urgent problem of theoretical investigations seems to be the search for specific unsteady-state solutions: oscillating, chaotic, etc., based on rheokinetic reasoning. One of the possible alternatives for initiation of oscillating states of motion for rheokinetic media is the sliding — adhesion transition which was discussed in Ref. [110].

Finally, significant progress could be achieved by establishing qualitative relations between the parameters of the process and the properties of the end product. This problem is beyond the scope of rheokinetic investigations because it requires one to elucidate the relations between molecular and supermolecular structure and the properties of the product. To this end, establishing the relations between the reaction pattern and "hydrodynamic" MMD, structure of macromolecules, etc. is required.

7 References

1. Stolin AM, Malkin AY, Merzhanov AG (1979) Polymer Eng. S. 19: 1065 and 1074
2. Malkin AY (1985) Plaste und Kautschuk 32: 281
3. Malkin AY (1985) Plastmassy 2: 29
4. Malkin AY (1987) In: Mashelkar RA, Mujumdar AS, Kamal R (eds) Transport phenomena in polymeric Systems — I., Wiley Eastern, New Delhi, (Adv. in Transport Processes, vol 5)
5. Malkin AY, Beghishev VP, Keapin IA, Bolgov SA (1984) Polymer Eng. S., 24: 1936
6. Malkin AY, Beghishev VP, Keapin IA, Andrianova ZS (1984) Polymer Eng. S., 24: 1402
7. Flory PY (1953) Principles of polymer chemistry, Cornell Univ. Press, Ithaca
8. Berlin AA, Volfson SA, Enikolopyan NS (1978) Kinetics of polymerization processes, Khimiya, Moscow (in Russian)
9. Lynn S, Huff JE (1971) AIChE 17: 475
10. Tadmor Z, Gogos CG (1979) Principles of polymer processing, Wiley, New York

11. Vinogradov GV, Malkin AY (1980) Rheology of polymers, Mir, Moscow/Springer, Berlin Heidelberg New York
12. Malkin AY (1980) Polym. Eng. S., 20: 1035
13. Malkin AY (1980) Usp. Khim. 50: 137
14. Budtov VP, Konsetov VV (1983) Heat transfer in polymerization systems, Khimiya, Leningrad (in Russian)
15. Stolin AM, Malkin AY, Merzhanov AG (1979) Usp. Khim. 48: 1492
16. Riesenberger JA, Sebastian DH (1983) Principles of polymerization engineering, Wiley, New York
17. Aris R (1983) Introduction to the analysis of chemical reactors, Prentice-Hall, New Jersey
18. Perlmutter DD (1965) Stability in chemical reactors, Prentice-Hall, New Jersey
19. Frank-Kamenetzkii DA (1967) Diffusion and heat transfer in chemical kinetics, Nauka, Moscow (in Russian)
20. Thomas RH, Walters K (1964) Quart. J. Mech. Appl. Math. 17: 39
21. Ide Y, White JL (1974) J. Appl. Polymer S. 18: 2997
22. Zhirkov PV, Estrin YI (1984) Polymer Process Eng 2: 219
23. Butakov AA, Maksimov EI (1973) Doklady AN SSSR 209: 643
24. Zhirkov PV, Bostandzhiyan SA, Boyarchenko VI (1980) Teor. osnovy khim. tekhnologii 14: 702
25. Stolin AM (1975) Fizika goreniya i vzryva, 425
26. Boyarchenko VI, Zhirkov PV, Yankov VI, Kernitski VI (1977) Heat Transfer — Soviet Research 9: 153
27. Boyarchenko VI, Zhirkov PV (1983) Teor. osnovy khim. tekhnologii 16: 66
28. Shastry JS, Fan LT, Erickson LE (1973) J. Appl. Polymer S. 17: 1339
29. Shrenk L, Kuester JL (1974) J. Appl. Polymer S. 18: 3109
30. Pearson JRA, Shah IT, Vieira ESA (1973) Chem. Eng. S. 28: 2079
31. Merzhanov AG, Stolin AM (1974) Zhurn. prikl. mekhaniki i tekhn. fiziki 1: 65
32. Merzhanov AG, Posetzelskii AP, Stolin AM, Shteinberg AS (1973) Doklady AN SSSR 210: 52
33. Vaganov DA, Zhirkov PV (1986) Chem. Eng. S. 41: 237
34. Moskovskii SL, Konsetov VV, Khokhlov VA, Spilchevskii IA (1974) In: Polimerizatsionnye protsetssy. Apparaturno-tekhnologicheskoye oformlenie i matematicheskoye modelirovanie (Polymerization Processes. Apparatus and Technological Setting and Mathematical Modelling) ONPO "Plastpolimer", Leningrad 174
35. Vaganov DA (1975) Zhurn. prikl. mekhaniki i tekhn. fiziki 168
36. Zhirkov PV, Bojarchenko VI (1987) Teor. osnovy khim. tekhnologii 21: 480
37. Knorr RS, O'Driscoll CF (1970) J. Appl. Polymer S. 14: 2683
38. Ahmad A, Treybig MN, Anthony RG (1977) J. Appl. Polymer S. 21: 2021
39. Schmidt AD, Ray WH (1981) Chem. Eng. S. 36: 1401
40. Hamer JW, Akramov IA, Ray WH (1981) Chem. Eng. S. 36: 1897
41. Vaganov DA, Boyarchenko VI (1984) In: Tezisy dokladov I Vsesoyuznogo simpoziuma po makroskopicheskoi kinetike i khim. gasodinamike (Reports of 1st All-Union Symposium on Macroscopic Kinetics and Chemical Gas Dynamics) Chernogolovka, 2, part I: 64
42. Jonsen GEH, Hoogstraten HW, Ouwerkork G (1981) Ind. Eng. Prd. 20: 177
43. Maksimov EI, Peregudov NI, Butakov AA (1975) Fizika goreniya i vzryva 4: 568
44. Wallis JPA, Ritter RA, Andre H (1975) AIChE 21: 686, 691
45. Lynn S (1977) AIChE 23: 387
46. Fahien RW, Stankovich I (1979) Chem. Eng. S. 34: 1350
47. Ahmed M, Fahien RW (1980) Chem. Eng. S. 35: 889
48. Ahmed M, Fahien RW (1980) Chem. Eng. S. 35: 897
49. Stolin AM, Khudyaev SI, Pribytkova KV, Lisovskaya TM (1981) In: Problemy tekhnologicheskogo goreniya (Problems of Technological Combustion) Chernogolovka 2: 130
50. Grishin AM, Nemirovskii VB, Panin VF (1977) Fizika goreniya i vzryva 1: 156
51. Sala R, Valz-Gris F, Zanderighi L (1974) Chem. Eng. S. 29: 2205
52. Vaganov DA (1977) Zhurn. prikl. mekhaniki i tekhn. fiziki 1: 114
53. Bostandzhiyan SA, Boyarchenko VI, Zhirkov PV, Zinenko ZA (1979) Zhurn. prikl. mekhaniki i tekhn. fiziki 1: 130
54. Malkin AY, Sherysheva LI, Kulichikhin SG, Zhirkov PV (1983) Polymer Eng. S. 23: 804
55. Grishin AM, Nemirovskii VB (1985) Zhurn. prikl. mekhaniki i tekhn. fiziki 1: 72

56. Vrentas JS, Huang WJ (1986) Chem. Eng. S. 41: 2041
57. Malkin AY, Zhirkov PV, Berezovskii AV, Enenstein GA, Bokareva AZ, Klochin AA (1984) Polymer Process Eng. 2: 207
58. Malkin AY, Enenstein GA, Berezovskii AV, Gordeev LS, Zhirkov PV, Klochin AA, Bokareva EZ, Nurmukhamedov SN (1986) Teor. osnovy khim. tekhnologii 20: 344
59. Vaganov DA (1984) Zhurn. prikl. mekhaniki i tekhn. fiziki 1: 96
60. Malkin AY, Lavochnik YB, Beghishev VP (1983) Polymer Process Eng. 1: 71
61. Malkin AYa, Lavochnik YuB, Beghishev VP (1984) Polymer Process Eng. 2: 27
62. Malkin AY, Lavochnik YB, Beghishev VP (1983) Vysokomolek. soedineniya 25 (A): 430
63. Malkin AY, Lavochnik YB, Beghishev VP (1984) Vysokomolek. soedineniya 26 (A): 775
64. Kolodzej P, Macocko CW, Ranz WE (1982) Polymer Eng. S. 22: 349
65. Davtyan SP, Zhirkov PV, Volfson SA (1984) Usp. Khim. 53: 251
66. Chechilo NM, Khvilivitskii RY, Enikolopyan NS (1972) Doklady AN SSSR 204: 1180
67. Davtyan SP, Surkov NF, Rozenberg BA, Enikolopyan NS (1977) Doklady AN SSSR 232: 379
68. Khanukaev BB, Kozhushner MA, Enikolopyan NS (1974) Doklady AN SSSR 247: 612
69. Manelis GB, Smirnov LP (1976) Fizika goreniya i vzryva 5: 665
70. Zhirkov PV, Bostandzhiyan SA, Boyarchenko VI (1979) In: Tezisy Vsesoyuznoi konferentsii "Matematicheskoe modelirovanie i apparaturnoe oformlenie polimerizatsionnykh protsessov" (Proceedings of the All-Union Conference "Mathematical Modelling and Apparatus Setting of Polymerization Processes") Vladimir: 56
71. Butakov AA, Zanin AM (1978) Fizika goreniya i vzryva, 1: 91
72. Vaganov DA, Zhirkov PV (1983) In: Tezisy dokladov VIII Vsesoyuznoi konferentsii po khimicheskim reaktoram (Reports of 8th All-Union Conference on Chemical Reactors) Chimkent 2: 371
73. Babadzhanyan AS, Volpert VA, Davtyan CP (1987) Doklady AN SSSR 293: 1155
74. Faulkner R (1985) Polymer Process Eng. 3: 113
75. Butakov AA, Shtessel' EA (1977) Doklady AN SSSR 237: 1422
76. Lindt JT (1982) Polymer Eng. S. 21: 424
77. Lindt JT (1983) Polymer Process Eng. 1: 37
78. Elbirli B, Lindt JT (1983) Polymer Process Eng. 1: 109
79. Janssen LPBM, Schaart BJ, Smith JM (1982) Proceedings of the 2nd Polymer Extrusion Conference, London, 15: 1
80. Werner R (1984) Polymer Eng. S. 23: 303
81. Stuber NP, Tirrel M (1985) Polymer Process Eng. 3: 71
82. Nangeroni JF, Eise K, Kidwell DS (1985) Polymer Process Eng. 3: 85
83. Malkin AY (1982) Plastmassy 4: 47
84. Malkin AY, Beghishev VP (1983) Polymer Process Eng. 1: 83
85. Malkin AY (1984) Mekhanika kompozitzionnykh materialov, 4: 362
86. Malkin AY, Beghishev VP (1982) Rheol. Acta 21: 629
87. Malkin AY, Shuvalova GI (1985) Vysokomolek. soedineniya 27 (B): 865
88. Tyabin NB, Dakhin OH, Gerasimenko VA, Baranov AV (1981) Mekhanika kompozitzionnykh materialov, 9: 1061
89. Tyabin NV, Dahin OH, Baranov AV (1982) Teplofizika vysokih temperatur 20: 122
90. Leonov AI, Shvartz AI (1972) Vysokomolek. soedineniya 14 (A): 695
91. Vaganov DA (1982) Zhurn. prikl. mekhaniki i tekhn. fiziki 1: 43
92. Castro JM, Lipshitz SD, Macosko CW (1982) AIChE 38: 973
93. Ryan ME (1984) Polymer Eng. S. 24: 686
94. Malkin AY (1985) Usp. Khim. 54: 548
95. Gřmela M (1984) Polymer Eng. S. 24: 673
96. Biesenberger JA, Gogos CG (1980) Polymer Eng. S. 20: 838
97. Sebastian DH, Biesenberger JA (1983) Polymer Process Eng. 1: 131
98. Lindt JT (1986) Polymer Process Eng. 4: 125
99. Castro JM, Macosko CW (1982) AIChE 28: 250
100. Domine JD, Gogos CG (1980) Polymer Eng. S. 20: 847
101. Debry HG, Charbonneaux TG, Macosko CW (1986) Polymer Process Eng. 4: 151
102. Rojas AJ, Adabbo HE, William RJJ (1981) Polymer Eng. S. 21: 634
103. Manzione LT (1981) Polymer Eng. S. 21: 1234

104. Kamal MB, Ryan ME (1980) Polymer Eng. S. 20: 868
105. Throne JL, Gianchandani J (1980) Polymer Eng. S. 20: 889
106. Nguyen LT, Suh NP (1985) Polymer Process Eng. 3: 37
107. Sandrell DJ, Macosko CW, Ranz WE (1985) Polymer Process Eng. 3: 57
108. Sebastian DH, Boukobbal S (1986) Polymer Process Eng. 4: 53
109. Sibal PW, Camargo RE, Macosko CW (1983–84) Polymer Process Eng. 1: 147
110. Vaganov DA, Zhirkov PV, Malkin AY (1987) Vysokomolek. soedineniya 29 (A): 412
111. Vaganov DA, Zhirkov PV, Malkin AY (1988) Progress and Trends in Rheology-II (Proc. Second Conf. Europ. Rheol., Prague 1986), Suppl. Rheol. Acta 26: 417
112. McGregor JF, Pendilis A, Hamielec AE (1984) Polymer Process Eng. 2: 179

Editor: K. Dušek
Received March 6, 1989

Crosslinking of Polyolefins

M. Lazár[a], R. Rado[b], J. Rychlý[a]

[a] Polymer Institute of Slovak Academy of Sciences, Bratislava, Dúbravská cesta
[b] Research Institute of Cables and Insulating Materials, Bratislava, Továrenská 14

This paper covers the knowledge of different procedures of polyalkene crosslinking. Besides the crosslinking, the effect of crosslink formation on the properties of polyolefins which may be of use in various applications have been dealt with.

Crosslinking methods based on the recombination of macroradicals and/or condensation of silanes are discussed, taking into account parallel side reactions which are decisive factors in the resulting crosslinking efficiency and network structure. From crosslinking pathways occurring through free radicals, radiolytic, photolytic and thermally induced generation of macroradicals and subsequent reactions of reactive intermediates are described in more detail.

Although much is already known in this field, the possibilities of its exploitation has not been fully explored yet. As an example, the idea of the recombination and disproportionation of low molecular alkyl radicals as an explanation of the crosslinking efficiency of polyolefins may be put forward. Of equal importance is to define the problems which await a more detailed study as well as to interpret sometimes controversial results and particularly to elaborate new principles of physical crosslinking of polyalkenes or chemically induced crosslinking of oriented macromolecules.

The paper uses the information from 198 references, of which one quarter were published in 1988.

1 Introduction

Contemporary methods of polyalkenes crosslinking result from long-term development originating at the outset of nuclear technology and applications of elementary particle accelerators. The utilization of these instruments made it necessary to quantify the effects of ionizing radiation on the materials used for constructing the respective equipment, including plastics. Research has shown that, in contrast to radiation disruption of solid crystalline substances, transformations may occur in organic polymers which are not always of a destructive character but that the structuralization of a polymer may take place, especially at lower irradiation doses. Relevant practical application of these investigations followed in the early 1950s, namely the crosslinking of saturated elastomers and polyethylene in order to modify their properties. Today, high energy radiation is widely used in the polymer industry and the corresponding technology has a multi-billion dollar market worldwide [1].

Because of the high costs of purposeful and regulated modification of polymers by irradiation, this route of crosslinking has expanded to large scale production only slowly. Simultaneously, more economical procedures of crosslinking have been looked for and various alternatives of chemically induced crosslinking were attempted, the most effective chemical agents found being peroxides. It has been established that a product can be obtained by chemical initiation of crosslinking which is similar to that obtained by ionizing irradiation. The possibility of using crosslinked polyethylene in high voltage insulation techniques became apparent and in the 1960s became industrially important and has been developed steadily ever since [2].

Attempts to use such a method of chemical modification in the production of polyethylene tubes which have no supporting elements and cannot be crosslinked above the melting temperature of polyethylene without an essential shape deformation led to the search for other more convenient methods. The most promising solution consists of the utilization of condensation reactions of silanes attached to a polymer chain which made it possible to design a new way of crosslinking with several technical and economical benefits. This so-called crosslinking through the silylation of a polymer was put into practice in the second half of the 1970s and found use in various applications of crosslinked polyethylene [3].

Significant progress has also been attained in two older routes of polyethylene crosslinking in which some technical problems, which had occurred in the initial period of application, were overcome. Three industrial technologies exist at present which differ from each other not only in the chemical principle of the processes but also in the effect attained. These technologies permit the selection of an optimum variant of crosslinking tailored to a given application of polyethylene. In the case of other polyalkenes, the amount of knowledge is considerably less.

2 The Chemistry of Crosslinking

2.1 General Principles

A chemical crosslink between polyolefin macromolecules can be formed by several alternative methods [4, 5] including the reactions:

a) Recombination of alkyl macroradicals

$$R^1\dot{C}H\,R^2 + R^3\dot{C}H\,R^4 \rightarrow (R^1)\,R^2 - CHCH - R^3(R^4) \tag{1}$$

b) Addition of a macroradical to a vinyl (vinylidene) bond attached to a polyolefin backbone

$$
\begin{array}{ll}
\text{\textcolor{white}{.}}\dot{C}H=CH_2 & \dot{C}H-CH_2R \\
& \\
\quad\quad\quad + R^{\cdot} \rightarrow & \\
& \\
CH_2=\dot{C}H & CH_2\dot{C}H^{\cdot}
\end{array}
\tag{2}
$$

Similar reactions may also occur on addition of polyfunctional low molecular monomers to a polymer system. For instance, during addition of a three functional monomer to alkyl macroradicals

$$R^1\dot{C}HR^2 + \overset{\|}{=\!\dot{M}\!=} \rightarrow R^1\dot{C}HR^2 \overset{=\dot{M}=}{\underset{|}{\text{\textcolor{white}{.}}}} \tag{3}$$

multiple bonds are incorporated into the macroradical which become the site of similar crosslinking reactions as depicted in Eq. (2).

Recombination of macroradicals, which are generated either chemically or physically, is one of the steps of the free radical mechanism of crosslinking. Thermally unstable compounds such as peroxides and azo compounds are usually used for free radical generation. The physical route of free radical generation in polyolefins includes mostly β-radiation (betatrones) and gamma rays (radioactive isotopes). In some cases, a combined effect of microwave excitation and a suitable peroxide, ultraviolet light, and a photosensitizer may be used.

c) Condensation reaction of reactive groups bound to a polymer chain.

In practice, condensation reactions of trialkoxy silane groups became significant.

$$
2\,\underset{\underset{R^2}{|}}{\overset{\overset{R^1}{|}}{CH}}-\underset{\underset{OX}{|}}{\overset{\overset{OX}{|}}{Si}}-OX + H_2O \rightarrow \underset{\underset{R}{|}}{\overset{\overset{R^1}{|}}{CH}}-\underset{\underset{OX}{|}}{\overset{\overset{OX}{|}}{Si}}-O-\underset{\underset{OX}{|}}{\overset{\overset{OX}{|}}{Si}}-\underset{\underset{R^2}{|}}{\overset{\overset{R^1}{|}}{CH}} + 2\,XOH \tag{4}
$$

In this reaction siloxane crosslinks are formed between polymer chains. Hydrolysis and condensation may involve also other alkoxy (OX) or hydroxy groups and a more complex crosslinked structure is formed.

d) Metathesis of cycloalkene side groups [6] represented by the following scheme:

$$
\left|\overset{(CH_2)_m CH}{\underset{(CH_2)_m CH}{\Big\langle}}\overset{\|}{} + \overset{CH-(CH_2)_n}{\underset{CH-(CH_2)_n}{\Big\rangle}}\overset{\|}{}\right| \rightleftharpoons \left|\overset{(CH_2)_m CH = CH(CH_2)_n}{\underset{(CH_2)_m CH = CH(CH_2)_n}{\Big\langle\Big\rangle}}\right| \tag{5}
$$

Although a crosslinking through metathesis has also been known for a fairly long time, it is used only for analytical determination of anomalous structural units on a polymer chain of polyolefins [7].

2.2 Disproportionation of Macroradicals

Recombination (dimerization) reaction of macroradicals may occur parallel to disproportionation in which alkyl radicals decay via transfer of a hydrogen atom from one β-carbon of one of the two radicals to the carbon of the second radical carying a radical site.

$$2 \text{ R}^1\dot{\text{C}}\text{HCH}_2\text{R}^2 \rightarrow \text{R}^1\text{CH}_2\text{CH}_2\text{R}^2 + \text{R}^1\text{CH} = \text{CHR}^2 \tag{6}$$

In such a way, one macroradical yields a saturated macromolecule, the other the macromolecule with end-chain unsaturation.

To establish the effect of structure on the ratio of recombination and disproportionation, the knowledge of reactions of low-molecular analogs may be of use. It may be deduced from the ratios of rate constants of disproportionation and combination determined for different types of alkyl radicals that primary alkyl radicals enter combination preferably while tertiary radicals disporportionate (Table 1). A relatively high proportion of disproportionation occurs also in the interaction of secondary alkyl radicals.

An increase of disproportionation at the expense of recombination when going from primary to secondary and tertiary alkyl radicals may be expressed by the ratio 1:5:25. An almost equal ratio for disproportionation and recombination has been found for reactions of the methyl radical with ethyl, isopropyl and *tert*.butyl radicals [13].

The proportion of disproportionation decreases on decreasing the number of β-hydrogens in the vicinity of a radical centre. However, the number of β-hydrogens available for the abstraction leading to disproportionation products cannot completely explain the k_d/k_c ratio found for alkyl radicals. The convincing objections against this simple and only partially valid rule are manifested in an example of disproportionation of unequal radicals [14]. Worth noticing is also a comparison with allyl radicals which enter combination preferably. The same declination can be seen on the k_d/k_c ratio for *tert*-butyl and α,α-dimethyl benzyl radicals in a liquid medium

Table 1. The ratio of rate constants for disproportionation (k_d) and combination (k_c) for different alkyl radicals in a gaseous phase

Radical	k_d/k_c	Ref.	Radical	k_d/k_c	Ref.
$\dot{\text{C}}\text{H}_2\text{CH}=\text{CH}_2$	0.008	8	$\text{CH}_3\dot{\text{C}}\text{HCH}=\text{CH}_2$	0.014	[8]
$\dot{\text{C}}\text{H}_2\text{CH}(\text{CH}_3)_2$	0.076	9	c-$\dot{\text{C}}\text{H}(\text{CH}_2)_5$	0.56	[11]
$\dot{\text{C}}\text{H}_2\text{CH}_3$	0.12	10	$\text{CH}_3\dot{\text{C}}\text{HCH}_3$	0.62	[12]
$\dot{\text{C}}\text{H}_2\text{CH}_2\text{CH}_3$	0.13	9	$\text{CH}_3\dot{\text{C}}\text{HC}_2\text{H}_5$	0.70	[9]
$\dot{\text{C}}\text{H}_2\text{CH}_2\text{CH}_2\text{CH}_3$	0.14	9	$\dot{\text{C}}(\text{CH}_3)_3$	2.5	[9]

[15]. While *tert*-butyl radicals disproportionate predominantly, α,α-dimethyl benzyl radicals recombine.

Before application of the above kinetic data for reactions of low molecular radicals to reactions of macroradicals in polyolefins, a difference between reaction conditions during determination of the values of the above rate constants and those in cross-linking should be accounted for.

The ratio k_d/k_c of rate constants for numerous alkyl radicals is relatively loosely dependent on temperature. The fact that some temperature dependence indeed exists indicates that transition states for disproportionation and combination cannot be exactly the same. The difference between the activated complex of the two reactions of radical decay may be confirmed from other experimental facts. Examining measurements of more than 40 papers of various authors on the decay reaction of ethyl radicals we find that values k_d/k_c determined for a liquid phase lie in the region 0.12–0.35. Although the results may be somewhat disturbed by experimental and methodical errors, the overall tendency of an increase of k_d/k_c with the increase of polarity of a solvent may be seen [10]. The lower limit of this ratio coincides with the gas phase value, the highest value of k_d/k_c was obtained for aqueous medium. It has been suggested that the transition state for combination is not greatly polarized while that for disproportionation has a polar character. The role of polar structure in the disproportionation reaction may be seen even in the case of relatively nonpolar alkyl radicals.

A considerably higher ratio of disproportionation occurs with more polar radicals such as primary or secondary alkoxy radicals and halogen alkyl radicals. The predominance of disproportionation over combination for polar radicals could be explained by electrostatic repulsion of the radicals with electric dipoles functioning against dimerization. The polar structure of radicals facilitates the proper orientation at the formation of an activated complex of disproportionation.

The effect of medium polarity on k_d/k_c ratio may be evident particularly in a condensed medium where interactions between macromolecules become more intimate. Therefore, for the same radicals, the values k_d/k_c found in a gaseous phase should be lower than those in a liquid or in a solid. This is in accordance with experiment (Table 2); unfortunately the number of k_d/k_c values determined for identical radicals in differing physical states of reaction medium is relatively low. It may be of interest that the k_d/k_c ratio also increases when going from the liquid into the solid state [18].

Table 2. The effect of a physical state of surrounding medium on the k_d/k_c ratio of rate constants of disproportionation and combination of radicals

Radical	State*	k_d/k_c	Reference
$\dot{C}(CH_3)_3$	g	2.5	[9]
	l	5.4	[9]
	s	20	[16]
c-$\dot{C}H(CH_2)_5$	l	0.6	[11]
s	1.1	17	

* g — gaseous, l — liquid, s — solid

A reduced mobility of reactants may cause other effects. The radicals in the solid phase can decay, e.g., in the sequence of fragmentation and combination reactions. Provided that a sufficiently mobile hydrogen atom is ejected from the radical primarily, it may combine with other radical and the final result will be identical with disproportionation. Thus when estimating the proportion of the competitive disproportionation reaction during crosslinking we may assume that *tert*.alkyl macroradicals will crosslink with a maximum efficiency five times lower than secondary macroalkyl radicals. Because of experimental conditions of crosslinking not being identical, these differences may be even higher. Any random branching of the linear structure of a polyolefin chain will reduce the degree of crosslinking brought about by dimerization of alkyl macroradicals. From this reason, the dimerization route is used in practice only for polyethylene. For efficient crosslinking of other polyolefins, other crosslinking reactions have to be used.

An addition reaction of polyalkene macroradicals to a multiple bond represents another crosslinking route which may become dominant in some cases. It includes two alternatives; a reaction with anomalous structural units of a polymer chain or a reaction of heterogeneous structural units deliberately attached to the polyolefin. The presence of a double bond in the polymer chain allows the formation of allyl macroradicals which promotes the efficiency of crosslinking in copolyalkenes [19, 20]. Incorporation of reactive functions to polyolefins enables one to conduct the crosslinking as a radical reaction even in the case of branched macromolecules which otherwise crosslink with low efficiency or even degrade.

2.3 Influence of Additives

Additives in polyolefins which improve their application properties affect the crosslinking whenever they take part in the generation of free radicals or enter the propagation reactions. With the presence of fillers with acid sites, peroxides are, e.g., decomposed by an ionic mechanism and formation of a crosslinked structure in polyethylene is suppressed. To remove such an effect, acid centres should be neutralized by basic substances.

A positive effect of fillers may be observed during irradiation crosslinking. It was found that the yield of radicals in polyethylene was increased by 50% when a small amount (0.05%) of aerosil was added [21]. It has been assumed that a higher production of radicals takes place at the interphase aerosil-polyethylene, where macromolecules can be in the nonequilibrium state of uncompensated strains. With a higher content of a filler, a transfer of energy from the filler to the polymer phase may occur and thus contribute to a higher yield of free radicals. Combination of irradiation with reactive admixtures may, moreover, affect a localization of crosslinks along the polymer chain.

Antioxidants are another group of additives which are necessary for any crosslinked mixture designed for the practical purpose of conferring a higher thermooxidative stability on a polymer product [22, 23]. Usually they affect the crosslinking negatively by scavenging radicals which may form crosslinks. An optimum choice of antioxidant thus depends on the purpose of the resulting product, on conditions of storage of a mixture to be crosslinked, etc. The amount of antioxidant for a required effect should be higher than that for a common thermo-oxidative stabilization of poly-ethylene. This is due to a higher operating temperature of a crosslinked polymer and to the necessity of overdosing the antioxidant because of its partial consumption during crosslinking. The overall amount of antioxidant used thus lies within a range of 0.2 to 0.8 % w., and in the presence of filler with a large surface, even higher.

Besides antioxidants, metal deactivators, light stabilizers, fillers, flame retardants, blowing agents, and pigments may be added to a crosslinked system. The right choice is conditioned by the inert behaviour of an additive towards the main reactants of the crosslinking process.

Reactive admixtures for free radical induced crosslinking include polyfunctional monomers; for an increase in the crosslinking efficiency of polyalkenes, trimethyl propane trimethacrylate, pentaerythritol tetraallyl ether, pentaerythritol tetrameth-acrylate, triallyl cyanurate [24, 25] etc. are used. An optimum concentration of poly-functional monomer depends on various factors and for a certain crosslinkable system can only be found empirically.

Polyfunctional monomers have been put to use in the crosslinking of polypropylene and of branched polyalkenes where polymer radicals decay in disproportionation reaction or fragmentate preferentially.

3 The Structure of the Network

Interchain bonds which arise through different pathways of crosslinking have different chemical structures. Combination of secondary alkyl macroradicals leads to tetra-functional carbon-carbon bond (H type crosslink), in which each carbon of a cross-link bridge binds two segments of the polymer chains. The real network, however, differs to a some extent from the simplified model (Fig. 1). In irradiated polyethylene, although they are less numerous, junctions of different functionality than in the H type are formed in conjunction with entanglements, and this represents a completely dif-ferent kind of network. Junction points of lower functionality appear during cross-linking due to specific structural units present in the system. After the combination of primary (chain-end) and secondary macroradicals, three functional (Y type of crosslink) junction points are formed while addition of a macroradical to the poly-functional monomer gives rise to polyfunctional junctions.

During crosslinking through silanes, polyfunctional network junctions are formed, too. From the silicon atom of a functional group four bonds emanate, one of them being directly, or via an ethylidene group, linked with the backbone of a polyalkene while the remaining three bonds with alkoxy groups are the potential sites of cross-linking. Attachment of a silane functional group to polyalkene is determined by the synthesis:

$$\begin{matrix} Si(OCH_3)_3 & CH_2CH_2Si(OCH_3)_3 & (7) \\ | & | \\ \sim CH_2CH_2CHCH_2\sim & R^1-CH-R^2 \end{matrix}$$

copolymer of ethylene   polyalkene grafted with vinyl trimethoxy silane

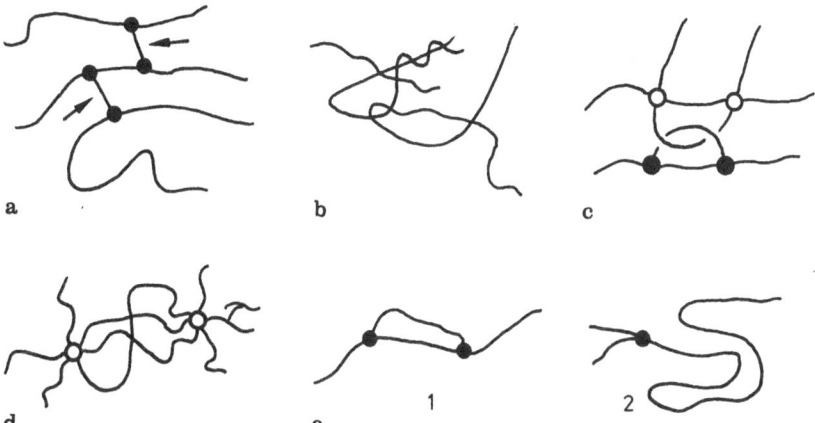

Fig. 1a–e. Schematic illustration of network structure: **a)** for a polymer with tetrafunctional C—C bonds (*arrows* indicate crosslinks).
b) physical entanglements of macromolecular chains
c) catenane-like entanglements of polymer chains; *full circles* represent three-functional branching points, *open circles* tetrafunctional branching points.
d) crosslinked structure with polyfunctional branching points
e) noneffective three functional branching points
1 — of intramolecular cycles
2 — at the chain ends

From kinetic and thermodynamic considerations it is assumed that not each of the three alkoxy groups of the silane will be capable of crosslinking [26]. On average, it occurs with two of them. One network junction thus formed binds several macromolecular chains (Fig. 2).

As far as the thermoelasticity performance of a polymer network is concerned, it is generally true that the higher the concentration of crosslinks or the lower the dimensions of loopholes in the network, the more significant are the changes in the polymer properties. The elasticity of a polymer is also enhanced by physically entangled chains, whose number increases with the number of crosslinks, the character of the crosslinks being interrelated with the character of the physical entanglements [27]. This may be illustrated by experiments with two crosslinked samples prepared from low density polyethylene. The first sample was crosslinked via the silane pathway, the second by

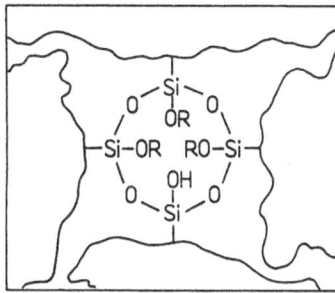

Fig. 2. Scheme of a polyfunctional branching point and of a part of network of a polyolefin crosslinked through the silane procedure

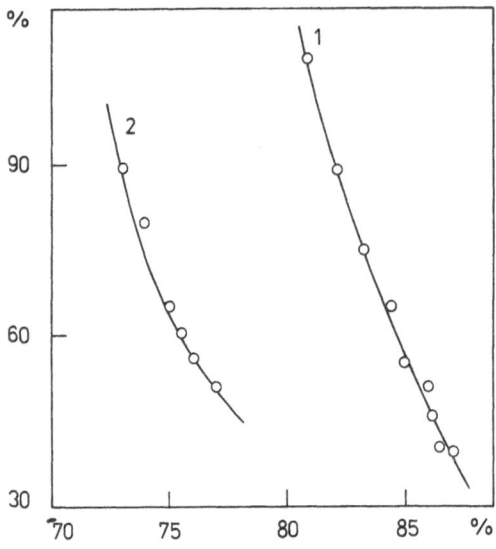

Fig. 3. Plot of relative elongation of samples loaded with 0.2 MPa at 200 °C on the content of a gel in low density polyethylene. Polyethylene was crosslinked by dicumyl peroxide (*1*) and vinyltrimethoxy silane (*2*)

peroxides only. The samples gave different results for thermomechanical resistance and solubility tests (Fig. 3).

Besides the chemical structure of a network junction, the properties of a crosslinked polymer are also affected by the distribution of crosslinks in the polymer matrix. The difference in properties of the polymer crosslinked by thermally labile compounds such as peroxides and/or ionization radiation may thus be evident.

Peroxide induced crosslinking is effected at temperatures when polymer is usually completely amorphous whereas crosslinking by γ-radiation is usually carried out when there are large crystalline regions in the polymer. Macromolecules crosslinked in an amorphous state form crystallites with difficulty on cooling; polyethylene crosslinked by peroxides therefore has a lower degree of crystallinity than that crosslinked by gamma rays irradiation with the same degree of crosslinking. The distribution of polyfunctional network junctions of the siloxane type is determined by the course of silylation and crosslinking, both considered separately. The attachment of silane groups, achieved either by copolymerization or by grafting, occurs so that these groups are situated in an amorphous phase. A relatively low concentration of silane side branches affects the crystallinity of a polymer only a little. Crosslinking of polyethylene is conducted here at temperatures of about 100 °C which may bring about the recrystallization and the increase of the crystallinity degree of the polymer [26].

It should be pointed out that the temperature of heating of a silylated polyethylene determines the maximum attainable concentration of crosslinks; at the same degree of silylation, the maximum concentration increases with increasing temperature parallel to an increase of the extent of an amorphous phase. This contributes to a higher proportion of silane branches available for the condensation reaction. Owing to the existence of crystalline regions, a nonhomogeneous network distribution arises during the crosslinking of either silylated or radiation treated polyethylene performed below its melting temperature. Distribution of a network density is a mirror image of dimensions and distribution of amorphous and crystalline regions throughout the polymer bulk.

The difference in crosslink distribution in the polymer network exists even in the case of crosslinking of polyethylene silylated by different procedures [28]. Shorter and sterically more compact branches of a statistical copolymer are less accessible by water and the process of intermolecular condensation occurs more slowly, when compared with grafted polyethylene. Besides the structural heterogeneity of silane side branches, different properties of silane crosslinked polyethylenes are due to differing distributions of the respective silane structural units which result from different types of silylation. Silane branches which are formed during a radical induced grafting of the ternary system of a polymer, silane, and peroxide can only be homogeneously distributed when peroxide and silane are homogeneously dispersed in the polymer [29].

The declination from an optimal regime causes crosslinking to occur in more or less homogeneous micelles of reacting components. Heterogeneties involving aggregates of functional groups undergo fast condensation due to a trace amount of water present in the polymer, and crosslinked regions thus appear in silylated products even during polymer processing. This is also a reason why polyethylene silylated by grafting is more difficult to process; the long storage of such a polymer may cause serious problems with its further processing. No ultimate recommendations can be given for the general or specific selection of a particular crosslinking procedure taking into account only the effect of the procedure on the character of the network structure and thus on functional and technological properties of a crosslinked polymer. The required extent of transformation of a thermoplastic polymer into that with thermoelastic properties can be attained with any method considered; it is sufficient to choose the proper temperature interval of polymer processing. The decisive technical criterion for the choice of a given crosslinking method results from the geometry and a shape of final product and from the technology of its production.

4 Measurements of Crosslink Density

The determination of the concentration of crosslinks in a crosslinked polymer sample is possible by several procedures [30]. The most frequent are: equilibrium swelling, stress-strain measurements and sol-gel analysis. These methods applied to identical crosslinked samples yield, however, different values of crosslink concentration [31, 32].

The highest values are attained in measurements of mechanical properties of the swollen polymer, lower in equilibrium swelling and the lowest in sol-gel analysis. The difference in results is mainly due to non-permanent physical entanglements, whose contribution decreases with a dose of irradiation to the polyethylene [33]. At high irradiation doses, the effect of chemical crosslinks is dominant and entanglement will occur only at high extension. In contrast, at low doses, the primary effect of irradiation is the stabilization of the natural entangled network by forming a small number of permanent chemical crosslinks. The gel fraction may still be very low, but the stabilized entanglements can provide a physical opposition to deformation resulting in a much higher stress than in the case of nonirradiated samples.

For all procedures of crosslinking, it was experimentally ascertained that the gel content depends on the molar mass of a polymer and on its distribution curve. The higher proportion of low molecular mass fragments in a polymer may, e.g., cause crosslinking to be less than 100 %. Most research studies on crosslinking have been conducted with commercial, i.e. nonfractionated polymers which might contribute to the quantitative disagreement of crosslinking efficiency observed by different authors.

Relations for molecular mass changes occurring during crosslinking of chains before the gel point are given in [34]. A methodology for computing network parameters such as the weight fraction of soluble material, the concentration of junction points, the weight fraction of a material that is a part of the network but elastically inactive and the entanglement trapping factor for various crosslinking models has been developed for various crosslinking models [35].

5 Crosslinking by Ionizing Irradiation

The use of radiation for polyethylene crosslinking, which has been known for several decades, belongs to the economically most successful products of radiation chemistry research [36, 37]. The main advantage of radiation initiation consists of the possibility of generating active intermediates in the solid polymer within a large temperature interval. The sources of radiation for the crosslinking of polyethylene industrially used are betatrones which allow one to obtain high radiation doses within a short time lag.

The primary step of interaction of radiation and polymer involves the formation of macroions

$$-CH_2- \rightsquigarrow -\overset{+}{C}H_2- + e^- \tag{8}$$

The electron released has so much energy that it may induce further ionization and excitation of other macromolecules.

The electron thus gradually transmits an excess of its kinetic energy to the environment and is finally captured by some cavity in the electroneutral polymer matrix

$$-CH_2- + e^- \rightarrow -\overset{(-)}{C}H_2- \tag{9}$$

or recombines with a positive ion. Recombination of particles of opposite charges leads to the formation of excited electroneutral macromolecules which are capable of cleaving to free radicals [38]. The dissociation may either involve an elimination of hydrogen atoms

$$-\overset{*}{C}H_2 - \rightarrow -\overset{\cdot}{C}H- + H\cdot \tag{10}$$

or cleavage of the carbon-carbon bond

$$-CH_2-\overset{*}{C}H_2-CH_2- \rightarrow -CH_2-\overset{\cdot}{C}H_2 + -\overset{\cdot}{C}H_2 \tag{11}$$

Part of the excitation energy may be transformed to the kinetic energy of dissociation products. This is a route by which hot hydrogen atoms of high reactivity are formed.

5.1 Chemical Transformations in Polyethylene

Although the C—H bond in polyethylene is stronger than the C—C bond, excited macromolecules eliminate hydrogen atoms preferentially through a C—C bond cleavage of a macromolecular backbone. This may be rationalized by two reasons. First, the dissipation of excitation energy along several identical C—C bonds linked to a sequence is easier than in the case of mutually isolated C—H bonds. The second reason is that the abstraction of relatively light hydrogen atoms contributes to the larger separation of primarily formed radicals than in the case of the formation of more bulky alkyl fragments. It is obvious that the more the radicals formed move away from the initial radical pair, the lower is the probability of their back self-recombination at the site of their generation and thus of regeneration of an original molecule.

Following C—H bond scission, the H atoms formed remain trapped in the vicinity of the alkyl radicals only at 4 K. A slight increase of temperature will induce their further reactions and decay. At somewhat higher temperature (e.g. at 77 K), disproportionation and transfer reactions occur within a radical pair of alkyl radicals and hydrogen atoms, and in ESR spectrum hydrogen atoms cannot be traced.

The reactivity of hydrogen atoms, within an initially formed radical pair may be illustrated by the fact that at 4 K about 50% of the radicals formed remain in the original radical pair, whereas at 77 K, it is only about 2%. Above 50 K, transiently formed primary alkyl radicals $-\dot{C}H_2$, abstract hydrogen atoms from the surrounding macromolecules and secondary alkyl macroradicals are predominantly formed. The other types of less reactive radicals may be detected on elevating either the temperature or irradiation dose. For instance, at a temperature of about 300 K, allyl

$$-CH_2-\dot{C}HCH=CHCH_2- \text{ and at higher temperatures polyene}$$
$$-CH_2(CH=CH)_n\dot{C}H- \text{ radicals are observed.}$$

On self-disproportionation of less numerous chain-end alkyl radicals, the chain-end double bonds and methyl groups are formed in polyethylene.

$$2-CH_2-\dot{C}H_2 \rightarrow -CH=CH_2 + CH_3CH_2- \tag{12}$$

The combination of chain-end radicals involves only a small fraction. A higher proportion of disproportionation may be expected with secondary alkyl radicals and hydrogen atoms.

$$-CH_2\dot{C}H-CH_2- + H^. \rightarrow -CH=CHCH_2- + H_2 \tag{13}$$

A reaction of hydrogen atoms with the CH_2 groups of a surrounding polymer chain and formation of secondary alkyl radicals and molecular hydrogen will compete with this reaction. Secondary alkyl radicals either recombine or disproportionate but the probability of both reactions is not so different now. Not only disproportionation of radicals but also their fragmentation may affect crosslinking negatively; macro-radicals cleave at the bonds of the main chain and molecular mass decreases.

5.1.1 The Yield of Crosslinking and Side Reactions

Radiation yield (number of molecules transformed or structural changes induced by absorption of energy of 100 ev) of formation of macroradicals [39, 40] is from 2 to 4, molecular hydrogen from 3 to 4, double bonds from 1.8 to 2, crosslinks from 1 to 2, chain-end methyl groups about 0.4, and main chain scissions from 0.2 to 0.7, respectively. Such a scatter of results was found under apparently identical experimental conditions.

The yield of internal double bonds is very much the same in various samples of poly-ethylene [40]. This should be reflected in a stationary concentration of intermediary products formed in the initial stages of irradiation. The concentrations of vinylenes in the crystalline and noncrystalline regions were found to be equal. A suggestion has been made that both *cis* and *trans* configurations were produced in the noncrystalline phase whereas only *trans* vinylenes appeared in the crystal. Concerning the methyl ends, concentration in the noncrystalline regions were approximately double those in the crystal.

It is known that increasing temperature increases the number of crosslinks formed as well as the amount of hydrogen, number of main chain scissions, and end-chain methyl groups, respectively. The concentration of unsaturated bonds is slightly re-duced which may indicate the participation of relatively unreactive vinylene groups in the polyaddition reaction occurring at higher temperatures, especially above melting temperature of polymer crystallites. An increase in the yield of hydrogen and cross-links may correspond to the higher yield of free radicals at higher temperatures and to the increase of free volume in the polymer system [41]. Reactive vinyl unsaturated bonds formed simultaneously will disappear in the polyaddition reaction. This was confirmed by high-resolution C^{13} NMR spectrometry measurements of irradiated polyethylene and by the model study of low molecular compounds [42, 43]. Vinyl unsaturated bonds will promote the radiation yield of insoluble gel which was experi-mentally verified on crosslinking of 1,2-polybutadiene where the yield of crosslinks was 14.3 [44].

The same kind of addition reaction to double bonds $C=C$ is probably a reason for an abnormally high radiation yield of crosslinks ($G=11$) observed during the first stages of polyethylene irradiation before the first fractions of an insoluble gel appear [45]. Since the scissions of the main chain may also occur as a parallel process, the crosslinking would have to be even higher in order to compensate it. The scissions of the polymer main chain are due to oxygen which is always present in trace quantities in polyethylene even in a high vacuum. A single molecule of oxygen is capable of breaking more than one macromolecule, the fact which cannot be explained by the only reaction of alkyl macroradicals with oxygen. A sensitizing effect of oxygen dis-solved in a polymer may consist in formation of oxygen anion-radicals and hydro-carbon cation-radicals which will fragment, etc. The oxygen anion-radical is trans-formed to a hydroperoxy radical which in a transfer reaction with the polymer gives a hydrocarbon macroradical and hydrogen peroxide. Provided that the hydrogen peroxide formed reacts with the polyethylene to give rise to two molecules of water, one molecule of oxygen may bring about the cleavage of four macromolecules. If a transfer reaction of primary alkyl radicals to the polymer and fragmentation of the secondary alkyl radicals are involved, the enhancing effect of oxygen on the main chain scissions induced by ionization irradiation may be even higher.

A decrease in the ratio of destruction and crosslinking reactions in polyethylene observed in the first instants of irradiation in an argon atmosphere was interpreted not only by the absence of sensitizing effect of oxygen in the sample but also by the dependence of the degradation reaction on changing molecular mass distribution [46]. This latter factor which is capable of changing the ratio of degradation and crosslinking appears to be less significant than the sensitizing effect of dissolved oxygen.

In the excess of oxygen, in air, definitely lower radiation yield of chain scissions of polyethylene was observed when compared with oxygen consumed [47]. G for scission of low density polyethylene reached the value 4.1, while G for oxygen consumption was 9.6.

Anomalous structural units existing in the polymer chain may be an additional source of a part of radicals formed. The proposal may be put forward that excitons could be trapped on some defects preferentially, free radicals being formed there with a higher radiation yield than on regular structural polymers. This assumption explains the higher yield of radicals and consequently the higher yield of crosslinks observed at the onset of irradiation. Main contribution to the high yield in this stage of reaction can originate only from the chain character of C—C bond formation including addition reactions to multiple carbon-carbon bonds which regenerate active radical centres for the formation of crosslinks.

5.1.2 Decay of Macroradicals

The kinetics of the recombination of polyethylene macroradicals is determined by the mobility of macromolecular chain segments and by chemical reactions which contribute to an approach of radical centers [48]. Chemical migration of macroradicals is mediated by transfer and fragmentation reactions occurring in the highest temperature region of macroradical existence. At lower temperatures, only the radicals of radical pairs decay; from radicals which are further away, only those which are situated within a reach of the motion of the macromolecular segments recombine. Reactions of macroradicals with low molecular mass radicals or atoms are made possible here by the diffusion of small molecules towards radical sites.

The decay of macroradicals is described by various schemes which usually hold for a certain interval of free radical concentrations. This situation arises probably because of the semicrystalline nature of polyethylene and because of a gradual change of mobility of macromolecules caused by crosslink formation. For radicals generated by radiation, the decrease of the rate constant of macroradical decay by about two orders of magnitude in polyethylene crosslinked by peroxides may be explained in the same way [49]. The steepest reduction in concentration of free radicals is observed as up to 70% of insoluble gel formation. The levelling of the rate constant of free radical decay, when the concentration of crosslinks is further increased, may be brought about by a simultaneously counteracting reduction of polyethylene crystallinity caused by crosslink formation. We know that reduction of polyethylene crystallinity increases the rate constant of free radical decay [50]. An increase in the proportion of crystalline regions retards radical decay in various types of polyethylene;

for identical specimens the decay depends on conditions of crystallization, which can be performed as a quenching, under pressure or in diluted solutions.

5.1.3 Effect of Crystallinity

The variations in a chemical structure of different types of polyethylene or ethylene copolymers [51] which follow from the character of production (such as low density polyethylene-LDPE, linear low density polyethylene — LLDPE, or high density polyethylene — HDPE) cause not only different radiation yields of crosslinks formation but also different radiation yields of other reactions [52]. For instance, in the range from 20 to 150 °C, LDPE has higher radiation yield of crosslinking than HDPE [53]. This is mainly due to the different degree of crystallinity of the polymers as well as to different sizes of lamellas in crystallites. The ratio of G value for crosslinking of crystalline and amorphous phase is about 1:2 for LDPE and 1:4 for HDPE.

On irradiation with the same radiation dose above and below the melting temperature of polyethylene crystallites [54], a higher gel fraction is formed with samples irradiated above the melting point.

To understand fully the differences in the structure of a semicrystalline polymer, the arrangement of the macromolecular chains in lamellas must be considered [55]. Polymer chains emerging from the face of a lamella into the region between crystals, which is normally 5–20 nm in thickness, pass through an interphase approx. 1.1 nm thick before reaching a fully disordered or isotropic state. In the interphase, approximately 70 % of the chains undergo a complete reversal, so that only 30 % enter the isotropic amorphous phase. Recombination of secondary alkyl macroradicals yields H type crosslinks in an amorphous phase, in which linked chains pass from one lamella to the other. The linkage of macromolecular segments of a chain on the phase interface has two alternatives. If a crosslink is formed on the fold of an identical macromolecule, a cycle arises involving several carbon atoms. Such small cycles affect the properties of polyethylene only slightly. On the other hand, during the linkage of such segments, entangled chains may be formed which will support the effect of crosslinks.

Increasing thickness of lamella or stretching macromolecular chains occurring during the crystallization of HDPE under high pressure leads to an increase of both allyl and alkyl macroradicals stability [56] (Fig. 4). In HDPE in which macromolecules are arranged into lamellas of thickness about 30 nm, isolated alkyl macroradicals decay below 330 K; if, however, macromolecules of polyethylene are stretched, the temperature of decay of corresponding radicals is about 80 K higher. Alkyl radicals in pairs decay in both morphological states with a comparable rate constant reaching the stability limit at about 100 K. Interesting results obtained for isolated radicals may be explained by a difference in the arrangement of macromolecular segments in noncrystalline regions and by a different distribution of amorphous islands in both examined samples.

The rare formation of crosslinks inside a crystal lattice is attributed to the fact that carbon atoms on adjacent chains are too far apart for interchain C—C bonds to be formed. Therefore, the lower the crystallinity of the same type of polyethylene, the higher the yield of crosslinks which are formed between macromolecules. The reduction in the lamella sizes leads to the increase of a bulk of interphase in a polymer

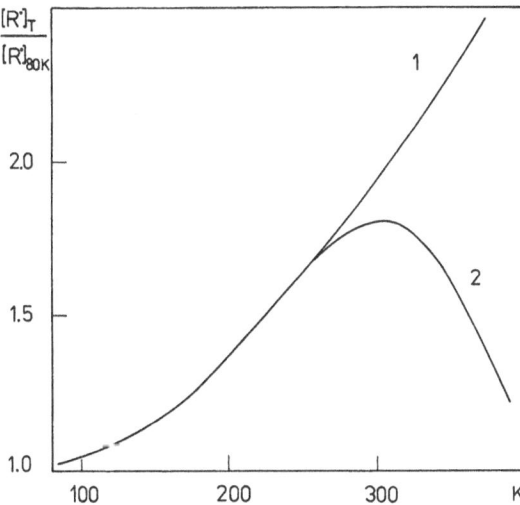

Fig. 4. Temperature dependence of the ratio of concentration of allyl radicals at the temperature T and 80 K for HDPE with stretched (*1*) and folded (*2*) macromolecules.

sample and the dose required to reach a gel point will thus decrease with decreasing temperature of polyethylene crystallization.

On the other hand, in arranged domains of macromolecules, polymer chains can be cleaved more easily than in amorphous regions. Such a conclusion ensues from the fact that after a disruption of a polymer chain, entropy of the system may rise more in an arranged macromolecular state than in random coils [57]. Since generation of radicals is more effective in an amorphous phase, the radiation transformation proceeding by a radical mechanism will occur there preferentially, and radiation destruction will be faster in oriented (strained) macromolecules of an amorphous state.

The crosslinks formed in less ordered regions at fold surfaces cause irregularities in the crystal lattice. Near the defect formed, a progression of further linking deeper down in the crystal on continued irradiation may be observed [58]. The result of irradiations is then a crosslinked polymer with a lower extent of crystalline regions.

The next crosslinks arise predominantly on an amorphous phase gradually formed from a crystal, as it was demonstrated on irradiation of crystalline paraffins [59]. In the light of the observed effects on crosslinking of polyethylene, the variations in values of radiation yields for respective changes of chemical structure as determined by various authors become more comprehensible. Regardless of experimental and methodical errors in determination of a given structural change, an important but insufficiently defined parameter is also the history of the polymer sample before irradiation. Sometimes only a minor declination in the processing regime of the same kind of polymer causes such a change in physical structure that it leads to a distinct scattering of results obtained.

The final effect of radiation will thus be determined by conditions to which a sample was exposed before and during irradiation and also by post-irradiation processes when reactive intermediates which did not succeed in reacting for some reason in previous stages will terminate the reaction. Another important factor may be the presence of impurities and admixtures in the polymer, which may either decrease or increase the yield of crosslinks.

5.1.4 Pathways to Increase of the Crosslinking Efficiency

The efficiency of radiation crosslinking may be increased by an addition of acetylene or other compounds with multiple carbon-carbon bonds [60]. The effect of acetylene may be explained by its addition to alkyl radicals and by the formation of allyl radicals which decay with a considerably higher ratio of combination and disproportionation than secondary alkyl radicals. Allyl radicals as side groups of macromolecules have, moreover, a higher degree of motional freedom. The effect of mobility of a side group in polyethylene is more pronounced below the melting point of polymer crystallites where the crosslinking efficiency may be five-fold higher. On the other hand, the efficiency of crosslinking above the melting temperature of polyethylene crystallites is approximately the same as without acetylene.

We may expect that polymers originating from monomers with several double bonds will give an increased yield of crosslinks on irradiation [61]. Besides consideration of the contact between reactive radicals one supplementary fact should be of importance, namely the residual number of double bonds in the polymer chain which may react further with primary alkyl macroradicals and thus increase the efficiency of crosslinking [62].

An interesting but not sufficiently understood effect is the synergism of ionization and ultraviolet radiation on polyethylene crosslinking. If, following the ionizing irradiation, polyethylene is exposed to ultraviolet light, the crosslinking efficiency will rise by a factor of 1.3. With the simultaneous effect of both radiation sources the efficiency is 5 times higher. Worth of noting is the observation that a yield of radicals does not increase noticeably. Gamma irradiated polyethylene releases hydrogen on exposure to ultraviolet light, which indicates the scission of $C-H$ bonds in the β-position to a radical site.

An increase of crosslinking efficiency with the simultaneous effects of UV light and gamma irradiation may be explained [63] by a tentative photosubstitution reaction of excited macroradicals

$$
\begin{array}{ccc}
-\dot{C}H- & & -CH- + H^{\cdot} \\
+ & \xrightarrow{h\nu} & | \\
-CH_2- & & -CH-
\end{array}
\tag{14}
$$

This scheme is simple and illustrative but not yet proved on model reactions of low molecular alkyl radicals [64]. Another alternative which takes into account the verified knowledge of photoisomerization of radicals, may involve the transformation of *tert*.alkyl radicals to secondary alkyl radicals. The increase of efficiency is then due to a more favourable ratio of k_d/k_c. Such an idea contradicts the relatively low proportion of branches even in a high pressure polyethylene. Obviously, the relay-like transfer reaction of a radical centre induced by ultraviolet light is more important here.

$$
-CH_2-CH- + h\nu \rightarrow -CH=CH- + H^{\cdot} \tag{15}
$$

$$
H^{\cdot} + -CH=CH- \rightarrow -CH_2\dot{C}H- \tag{16}
$$

$$
H^{\cdot} + -CH_2- \rightarrow -\dot{C}H- + H_2 \tag{17}
$$

$$
H_2 + -\dot{C}H- \rightarrow -CH_2- + H^{\cdot} \tag{18}
$$

This sequence of reactions ensures easier migration of radical sites each to other. An increase in the rate of collisions of macroradicals in a relatively rigid matrix will reduce the probability of main chain scissions. Double $C=C$ bonds will contribute to the increase of efficiency not only by the addition reactions but also by a possibility of creating macroradicals which terminate by combination exclusively.

Another significant fact is the appearance of ethylene with the parallel effect of UV light on gamma irradiated polyethylene [63] which may enhance the formation of bridges between relatively imobile macroradicals in a polyaddition reaction.

Taking into consideration the significant effect of ultraviolet light on polyethylene crosslinking, as well as the fact that irradiated polymers are capable of emiting light in a transformation of absorbed energy of ionization irradiation [65], pure radiolysis of polymers may involve photochemical reactions of macroradicals as one of the routes of polymer transformation. The effect of radiation on reactive intermediates need not necessarily lead to an increase of crosslinking efficiency. When using fast electrons for the generation of free radicals, it was established [66] that with the same overall dose but with a 10 times higher intensity of β-radiation, the number of crosslinks in LDPE is reduced at the expense of a higher yield of macroradical fragmentation. Under such conditions macroradicals dehydrogenate and disproportionate to a higher extent than at lower intensities of β-radiation.

5.2 Structural Changes in Polypropylene

When comparing the effect of irradiation on polyethylene and polypropylene an interest may be evoked by one order lower radiation yields of crosslinking and chain scissions for polypropylene. The G value for polypropylene irradiated in vacuum usually lies within the range 0.1–0.3 for chain scission and 0.07–0.18 for crosslinking [45]. Lower values of these yields can be associated with higher degree of crystallinity of isotactic polypropylene than that of polyethylene. At the same time, splitting off of side methyl groups from polypropylene occurring within a cage of surrounding medium is more difficult since a back recombination in a radical pair $(R^{\cdot} \; {}^{\cdot}CH_3)$ is more probable than a similar reaction of a mobile hydrogen atom and alkyl radical in polyethylene. This corresponds with the reduction of a radiation yield of molecular hydrogen formation to 75 % of the value attained for polyethylene. It may be of interest that this value is considerably higher than that for crosslinks formation which could indicate that, in polypropylene, intramolecular reactions of hydrogen atoms predominate over intermolecular reactions. Obviously, disproportionation reactions of hydrogen atoms with alkyl radicals will be more important than the transfer of hydrogen atoms to surrounding macromolecules.

5.2.1 Loss of Tacticity

Existence of cage reactions between macroradicals and low molecular fragments formed in radiolysis is increased by a change of stereoregularity of polypropylene occurring during irradiation [68]. The reaction, accompanied by the change of tacticity, has a considerably higher radiation yield than crosslinking or any other reaction. For the same total dose, the radiation yield of the conversion in stereoregularity is greater on irradiation with a higher dose rate. The greater density of an energy input will evidently reduce the ratio of uneffective dissipation of energy to vibrational degrees.

A loss of isotactic pentade sequences in a polypropylene film with prevailing isotactic units when subject to a electron beam (dose rate 720 kGy/h, 1.3 MeV) occurs with a G value of 220, i.e. 2.3 times higher value than the G value [94] for an identical sample irradiated with gamma radiation from a Co^{60} source (dose rate 4.4 kGy/h). A similar dependence on dose rate was observed with nonisotactic pentade sequences which appear during irradiation. The change in tacticity is probably caused by an inversion of the configuration on a radical site of asymmetric carbon atoms. Isomerization of polypropylene may occur after cleavage of any bond in the vicinity of the asymmetric carbon followed by an inversion of alkyl radical and recombination within a radical pair. The energy released in the recombination reaction contributes to the same reactions taking place nearby which is a reason for the relatively high yields of stereoregularity change. The reduction of isotacticity and presence of destruction reactions in irradiated polypropylene lead to a substantial decrease of crystallinity and of melting temperature. For instance, the decrement of melting temperature of polypropylene crystallites is about 100 °C after irradiation with a dose of 18 MGy [69].

5.2.2 Disproportionation and Fragmentation

The ratio of yields for chain scissions and crosslinking in identical polypropylene samples is about 1.5. The higher ratio of main chain scissions in polypropylene compared with polyethylene has several reasons. Back recombination of macroradicals formed during main chain scissions has a lower probability in polypropylene than in polyethylene. For primary alkyl radicals of polyethylene $k_d/k_c = 0.1$, while in polypropylene where a reaction of secondary and primary radicals may take place, k_d/k_c attains the value 0.3 [70]. The decisive factor governing the crosslinking is the fate of macroradicals formed after elimination of side substituents of a polymer molecule. If a small number of branches and chain-ends groups on which macroradicals may also be formed, is neglected, polyethylene gives rise to secondary macroradicals predominantly with the $k_d/k_c = 0.7$. The more complicated pattern arises with polypropylene in which practically all types of alkyl radicals may be transiently formed. With regard to the probability of possible alternatives of transfer reactions of macroradicals, *tert*.alkyl macroradicals [71] ($k_d/k_c = 2.5$) will be the most numerous in the system. Their termination reactions however, will lead to a reduction of crosslinking efficiency because of an increased amount of disproportionation.

Primary radicals with a radical site situated on pendant methyl groups will be of use for crosslinking only in a small extent. Tertiary alkyl radicals will arise in the system mainly as a product of transfer reactions of more reactive radicals. Reaction of methyl radicals with the tertiary hydrogen of polypropylene macromolecules is an example. Compared with a transfer reaction of a hydrogen atom in polyethylene, a methyl radical is less reactive and reaction thus may take place at a larger distance from the place of radical generation. The radicals of polypropylene will be further away from each other than in polyethylene and their mutual interaction to give rise a crosslink will occur with lower probability at the expense of increased probability of side reactions.

If irradiation of polypropylene is carried out in the presence of acetylene, the efficiency of crosslinking increases substantially [72]. This may be explained by a reaction of alkyl macroradicals with acetylene to give radicals of the allyl type which are known to combine preferentially.

Another fact which manifests itself in lower efficiency of the radiation crosslinking of polypropylene than that of polyethylene consists of lower activation energy (by 15 kJ/mol approximately) of the fragmentation of *tert*.alkyl radicals compared with secondary alkyl radicals. This difference corresponds to an approximate 500 fold higher rate constant of fragmentation for *tert*.alkyl radicals at room temperature. The higher activation energy of β-fragmentation than that of the recombination of alkyl macroradicals indicates that the efficiency of crosslinking decreases with increasing temperature. The elementary act of recombination of alkyl macroradicals depends also on a facility of mutual approach of radical sites, which, in contrast, is promoted by increasing temperature. Hence, a plot of crosslinking efficiency on temperature can display a maximum which was indeed experimentally confirmed for photosensitized crosslinking of polypropylene [73].

5.2.3 Kinetics of Macroradical Decay

Rate constants of macroradical decay in isotactic polypropylene change in relation to the content of the crosslinks [74]. For the first stage of crosslink formation, the rate constant of radical decay decreases by about 1 order; for higher conversions of crosslinking the rate constant increases. The initial decrease of the rate constant seems to be associated with a reduced mobility of macromolecular segments, while the subsequent increase with a gradual reduction of polymer crystallinity.

The macroradical decay curves exhibit two reactivity regions, one at about 170 K (region I) and the other at about 250 K (region II). The molecular mechanism for the approach of the radical centers was tested from the viewpoint of their physical and/or chemical migration [75]. Two mechanisms were suggested for the mutual approach of the centers in macroradical decay regions I and II in the amorphous zones of the matrix. Both involve the contribution of segmental conformational motions, the mobility of the chain-end segments being of importance in the region I whereas in the region II the mutual nearing of the reactants proceeds in a relay-like way via intermolecular abstractions of hydrogen atoms in conjunction with crank and crankshaft motions of the chain-end and internal segments of the chain.

5.2.4 Polyfunctional Monomers

Low efficiency of the radiation crosslinking of polypropylene is a reason why irradiation cannot be used for improving the thermomechanical properties of this polymer as it is commonly practiced with polyethylene. On the other hand, this fact has stimulated a large amount of research on polyfunctional monomers aimed at a deliberate increase of radiation efficiency of polypropylene crosslinking [76].

If butadiene is used as a coagent, the dose for gel formation is reduced by a factor 5 [77]. It is of interest that parallel to an increase of radiation yield of crosslinking, the yield of chain scissions increases, as well. For instance, for butadiene, G of crosslinking is 0.69 and G of scissions is 0.44; for divinyl sulphone ($CH_2=CHSO_2CH=CH_2$) this value is 0.46 and 1.1, respectively. From the G value it may thus be seen that the polyfunctional monomer reduces the ratio of cage recombination within radical pairs and converts the radicals to those which have more favourable ratio k_d/k_c for crosslinking. A different effect of monomers on G of main chain scissions demonstrates simultaneously a possible participation of ionic mechanisms in the radiation induced degradation of polymers [78].

When a polymer is irradiated below the melting temperature of crystallites, the polyfunctional monomer remains in the amorphous regions. Gels of polypropylene crosslinked in the presence of butadiene are crystalline which is a manifestation of the formation of crosslinks at the lamellar surfaces [79]. The presence of hexadecane and hexadecene-1 in isotactic polypropylene increases the rate of termination of radicals formed in the irradiated polymer and the oxidative post-effect is reduced as well as the deterioration of mechanical properties is less than in a polymer without additives [80].

6 Photocrosslinking

A direct irradiation of pure polyolefins by ultraviolet light practically does not initiate the crosslinking for which the absorption of light of very short wave-lengths is necessary. Impurities and low fractions of carbonyl groups which are present in commercial polyolefins, however, shift the maximum absorption to longer wave-lengths of about 280 nm [81]. Following the absorption of a photon, the carbonyl group becomes excited and may undergo the reaction

$$-CH_2COCH_2- \xrightarrow{h\nu} 2\ R\dot{C}H_2 + CO \tag{19}$$

and alkyl radicals formed may subsequently produce crosslinks in an inert atmosphere. The light also promotes the abstraction of hydrogen atoms from alkyl macroradicals and the formation of *trans* vinylene unsaturated bonds. Vinyl unsaturated bonds disappear during crosslinking. The variations in the concentrations of vinyl and vinylidene unsaturated bonds which are observed during photocrosslinking [82] may be interpreted as a chemical transformation of C=C bonds accompanied simultaneously by a formation of crosslinks as follows from Eq. 20.

$$
\begin{array}{ccc}
-CH{=}CH_2 & & -\underset{|}{C}{=}CH_2 \\
& \rightarrow & \\
-CH_2{-}CH_2{-} & & -CH{-}CH_2{-} + H_2
\end{array}
\tag{20}
$$

Within this scheme, however, the transitory formation of vinylene macroradicals is questionable.

Unsaturated bonds in polyethylene are not the product of the photofragmentation of macroradicals alone but also of their disproportionation. The disappearance of C=C bonds corresponds to addition reactions of radicals with multiple carbon-carbon bonds. Addition reactions increase the quantum yield of crosslinks when related to that of formation of free radicals.

6.1 Photoinitiators

On addition of a photosensitizer or photoinitiator to a polyolefin, the efficiency of photocrosslinking increases considerably. As photoinitiators, S_2Cl_2, $SOCl_2$, PCl and other compounds may be used which decompose into free radicals by the effect of ultraviolet light [83]. To attain a sufficient rate of crosslinking, a relatively high initial concentration (above 5% w.) of the above inorganic chlorides should be used. Another disadvantage of these photoinitiators consists in their toxicity and corrosion effects. This latter aspect is less distinct when using chlorinated hydrocarbons; here, however, the efficiency of crosslinking is reduced.

Photolytic decomposition of peroxides is not very efficient in crosslinking. An enhancement effect on the extent of photocrosslinking of polyolefins in the presence of peroxides is displayed by aromatic hydrocarbons such as naphthalene. These transfer the excitation energy absorbed to a peroxide. This procedure, however, does not represent an important improvement when compared with that referred to earlier, namely the photoreduction of polyethylene with aromatic ketones and quinones [84]. From aromatic ketones and quinones, particularly benzophenone [32], chlorinated benzophenones, benzoyl-1-cyclohexanol [82], α,α-dimethoxy-phenyl acetophenone, 2,4,6-trimethyl benzoyl phenyl phosphinic ethyl ester [85], anthrone [86], anthraquinone [87], naphthoquinone, benzoquinone, and their derivatives have all been examined.

Ketones and quinones, in which the lowest triplet state undergoes $n\pi$ transition, photoreact with polyethylene so that a radical pair is initially formed on irradiation [88].

$$R^1CH_2R^2 + O{=}CR^3(R^4) \rightarrow R^1\dot{C}HR^2HO\dot{C}R^3(R^4) \tag{21}$$

If the radicals of this radical pair move away, the combination of macroradicals may occur and crosslinks in polyethylene may be formed. Self-reaction within a parent radical pair remains, however, still more probable and reduces the yield of crosslinks.

6.2 Efficiency of the Process

Photoreactions of aromatic ketones and quinones take place both in vacuum and in air. The subsequent reactions of alkyl radicals with oxygen significantly affect crosslinking since parent ketones are regenerated from ketyl radicals and simultaneously, hydrogen peroxyl radicals are formed.

$$HO\dot{C}R^3(R^4) + O_2 \rightarrow O{=}CR^3(R^4) + HO_2 \tag{22}$$

These radicals can abstract hydrogen atoms from the polyethylene chains and give rise to new polymer alkyl radicals. This contributes to the increase of crosslinking efficiency. The train of subsequent reactions of radicals leads, moreover, to the appearance of a new precursor of free radicals, hydrogen peroxide. On the other hand, macroradicals of polyethylene may react with oxygen directly to form peroxy radicals which will reduce the yield of carbon-carbon crosslinks.

The efficiency of crosslinking obviously depends on the accessibility of oxygen to a radical pair. Small concentrations of oxygen in polyethylene will increase the efficiency of crosslinking when compared with a strictly inert atmosphere but an excess of oxygen will reduce it below that in vacuum. The accessibility of radical pairs towards oxygen is controlled by the thickness of an irradiated film, by the intensity of light, by the concentration of oxygen in surrounding atmosphere and temperature. For the case of polyethylene films irradiated in the air and in the presence of aromatic ketones (in the temperature range 20 to 60 °C), the yield of crosslinks is usually higher than that in vacuum.

Within the range from −40 to +60 °C the crosslinking efficiency increases with temperature at which irradiation has been performed. This is in agreement with a decrease of a maximum degree of crosslinking with the decrease of temperature. Above 60 to 70 °C, the amount of the gel fraction starts to decrease while below −40 °C, no photoinitiated crosslinking or oxidation of polyethylene was observed. The photocrosslinking may occur only above the glass transition temperature of the polymer. The increase of efficiency with increasing temperature above T_g corresponds to the diminuition of the importance of reactions within a caged radical pair. At higher temperatures (70 to 80 °C), when the temperature of λ-transition has been surpassed, and the motion of macromolecular aggregates in an amorphous phase is released, the diffusion of oxygen into the polymer bulk is enhanced [89]. The reaction of macroradicals with oxygen then causes a decrease in photocrosslinking efficiency.

Photocrosslinking of films requires 0.1 to 1 % w. of sensitizer. The use of higher concentrations of aromatic ketones or quinones is not recommended because the efficiency of crosslinking is reduced and crosslink distribution becomes nonhomogeneous, the highest density of crosslinks being accumulated in the upper photoexposed parts of the film. The photoexposition times are about 20 s at a light intensity of 10^{17} Einstein $\cdot$ cm^{-2} $\cdot$ s^{-1}. This enables one to perform photocrosslinking of polyethylene films in a simple continuous process. An addition of 1 % by w. of triallylcyanurate shortens the time of irradiation of different polyethylenes almost to a half of the original value.

To obtain a higher efficiency of photocrosslinking and to improve technological properties of prepared films (smoothness of the surface, transparency), 3 to 5 % of ethylene-vinyl acetate copolymer having about 12 % of vinyl acetate groups are added [90]. Increase of the molecular mass of the added copolymer leads to the increase of insoluble gel which indicates that this copolymer incorporates into the polyethylene network. The decrease or increase of vinyl acetate content in the copolymer below/above an optimum value always reduces the efficiency of photocrosslinking.

The reduction in the amount of a gel formed in a copolymer with a lower content of vinyl acetate structural units is connected with the lower sensitivity of the system to ultraviolet light. The higher content of acetate groups promotes the initiation of degradation reactions. Etylene-vinyl acetate copolymers admixtures contribute to the increase of photocrosslinking efficiency also by an increase of the extent of amorphous structures and by a formation of finer spherulitic structure in polyethylene. A similar effect has also been observed during peroxidic or radiation initiated crosslinking.

From other polyolefins, photocrosslinking of ethylene propylene elastomers, copolymerized with dienes such as dicyclopentadiene, 1.4-hexadiene, and 5-ethylidene

2-norbornene should also be quoted [82]. Crosslinking starts with the abstraction of the labile hydrogen atom which is in the β-position to noncopolymerized C=C unsaturated bonds originating from diene. Macroradicals of the allyl type are thus produced and recombine to yield crosslinks.

A further positive impact on photocrosslinking efficiency may be expected on the parallel application of magnetic field, as it was reported, e.g. for photocrosslinking of butadiene copolymers [91].

7 Crosslinking by Thermoreactive Compounds

7.1 Peroxides as the Source of Radicals

Peroxides initiate the crosslinking of polyolefins by a homolytic cleavage of the O—O bond to alkoxy radicals

$$ROOR' \rightarrow RO^{\cdot} + R'O^{\cdot} \tag{23}$$

which abstract hydrogen from the surrounding molecules

$$RO^{\cdot} + -CH_2CH_2CH_2 - \rightarrow ROH + -CH_2\dot{C}H-CH_2- \tag{24}$$

and produce alkyl macroradicals which recombine to form a crosslink. Homolytic cleavage of peroxidic bonds is affected by the structure of peroxide and by polar admixtures in polyolefins. The choice of peroxide for crosslinking of a polymer depends particularly on the thermal stability of peroxide related to the temperature of polymer processing. In technological practice other demands on the properties of peroxides used are met, namely that of handling safety, low toxicity, the route of dosing, etc. and only a few peroxides comply with all these requirements [92–95] (Table 3). The disadvantage of dicumyl peroxide, which is used in applications in polyethylene crosslinking up to now, consists in a parallel formation of acetophenone which has an offensive odour.

In the presence of polymer crystallites, the reaction of peroxide decomposition occurs with a rate about 2 to 3 times lower than in low molecular solvents [96], while

Table 3. Parameters of Arrhenius equation determined from decomposition rate constants k of some peroxides in nonpolar low molecular medium and calculated temperature T of peroxide decomposition related to the half-life 1 minute

Peroxide	A min^{-1}	E kJ/mol	T °C
2,5-Dimethyl-2,5-di-*tert*.butyl peroxyhexane	1.2×10^{14}	124	182
Tert.butyl cumyl peroxide	8.4×10^{12}	115	186
Ditert.butyl peroxide	5.4×10^{14}	131	187
2,5-Dimethyl-2,5-di*tert*.butyl peroxyhexin	1.1×10^{15}	137	198

in a polymer melt the difference is not so distinct. A slight deceleration of the peroxide decomposition in the polymer melt is due to the back recombination of alkoxyl radicals which causes the partial regeneration of a parent peroxide. However, primary radicals may self-react before escaping from the reaction cage without a restoration of the peroxidic bond. This case together with reactions of low molecular radicals and macroradicals formed, will reduce the efficiency of peroxides in polymer crosslinking.

The presence of polymer crystallites increases the ratio of back recombination of alkoxyl radicals at the expense of other competition reactions.

Peroxidic initiator is compounded with polyolefins either pure or blended with an inorganic carrier, or as a concentrate in the polymer. The addition of peroxide is carried out in a processing stage before thermal shaping of the polymer to the final product. Decomposition of peroxide is usually accomplished by an increase of temperature in an inert atmosphere. Water vapour which was originally used as a heating medium in crosslinking appeared not to be suitable for high voltage cable insulations since water penetrates to polyethylene and forms there a large amount of pores and cavities of magnitude about 20 μm [97]. Such defects may negatively affect the operation safety of cable conduit lines. Instead, nitrogen [98], silicon oil [99] and fused eutectic mixtures of salts [100] are used.

In some cases, polymer crosslinking may be effected by a post-addition of peroxide to the shaped polymer. The process is performed by sorption of peroxide from the surface of the extruded tube following its deposition [101]. Conditions of crosslinking are adjusted so that diffusion into the bulk of polymer precedes the decomposition of peroxide. The rate of diffusion depends on the kind of peroxide, used and for the same peroxide also on the type of polyethylene and its thermal history [102].

7.2 Efficiency of Peroxide Crosslinking

The number of crosslinks formed in a polymer may be expressed from the amount of decomposed peroxide and from crosslinking efficiency. If the reaction conditions are chosen so that about 5 half-lives of peroxide decomposition are reached, then the amount of undecomposed peroxide (about 3% of its original concentration) may be ignored.

A number of crosslinks formed in a macromolecular system related to the decomposed peroxide depends not only on the structure of the polymer chain but also on the source of primary radicals. For crosslinking of a polymer, it is necessary that the transfer reaction of oxy radicals predominates over the competitive fragmentation. If the two macroradicals are formed at close proximity, they may give rise to a one crosslink in the mutual recombination with high probability. The less favourable situation arises when oxyradicals undergo parallel fragmentation. Since carbon centered radicals are less reactive in the transfer reaction with polyethylene than oxy radicals, the probability of crosslinking is reduced.

For the decomposition of dicumyl peroxide in LDPE (2% w.) at 160 °C, the efficiency of peroxide in crosslinking was about 87% [103]. Such a high efficiency may be ascribed to the contribution of physical entanglements to the overall number of chemical crosslinks and to the presence of unsaturated C=C bonds in a polymer. We may recall that during dimerization of low molecular alkanes initiated by *tert*.butyl perbenzoate, an efficiency of only 50% was found [104]. One third of the primary radicals are consumed in the formation of double bonds in alkane, the next 15% of low molecular mass radicals derived from peroxide self-decay and about 5% combine with the alkyl radicals of alkane.

Secondary radicals in alkane do not fragment to a measurable extent, which indicates that fragmentation reactions in polyethylene will occur only at anomalous structural units, such as points of branching, etc. The presence of such structural defects reduces the efficiency of crosslinking. When compared with secondary radicals, tertiary alkyl radicals originating from these branching points undergo fragmentation rather than disproportionation. Polypropylene, when compared with polyethylene, may be therefore crosslinked only by a 20 fold higher concentration of peroxide [105].

7.3 Effect of High Pressure

Besides the temperature and medium quality, the rate constant of the decomposition of peroxide is affected also by pressure. Increasing pressure leads to a reduction of the sizes of empty intermolecular cavities in a polymer. The number of free gaps which have sizes corresponding to activation volume decreases. These gaps are necessary for a required separation of substituted oxy radicals formed from an O—O bond in the activated complex. At a pressure of 800 MPa, the rate of decomposition of, e.g. benzoyl peroxide is reduced by about 100 times. The activation volume determined from the plot of rate constant on pressure is 9.1 $cm^3 \ mol^{-1}$. Parallel to deceleration of benzoyl peroxide decomposition by pressure, the crosslinking efficiency also decreases. Comparing the number of crosslinks estimated from steady parts of the dependence of the gel amount on time for 3% w. of initially present dibenzoyl peroxide, we may find that at 800 MPa the crosslinking efficiency is only 65% of that obtained at 0.5 MPa [106]. This lowered efficiency seems to be due to side reactions of formed radicals.

The effect of pressure on the rate of decomposition of peroxide may have some positive impact on technological procedure [107].

7.4 Redox Initiation

Post-heating of a shaped polyethylene which induces its crosslinking may sometimes lead to an undesired deformation of the product. To overcome this difficulty, two-component redox-initiating systems producing free radicals at lower temperatures have been designed [108]. The reaction system involved cumyl hydroperoxide and a transition metal ion, whose higher and lower oxidation states differ by one electron. The decomposition of hydroperoxide proceeds by an electron transfer mechanism; an electron is transferred from the metallic ion (e.g. Co^{2+}) to the peroxidic bond which splits into two fragments.

$$ROOH + Co^{2+} \rightarrow RO^{\cdot} + {}^{-}OH + Co^{3+} \qquad (25)$$

The Co^{3+} ion formed is reduced back by hydroperoxide to the original oxidation state (Co^{2+}).

$$ROOH + Co^{3+} \rightarrow RO_2^{\cdot} + H^+ + Co^{2+} \qquad (26)$$

The sequence of these two reaction steps is repeated many times even at initial concentrations of Co^{2+} ions which are 1000 times lower than that which corresponds to a molar concentration of hydroperoxide. As the above reaction cycles of hydroperoxide decomposition yield only one radical per one molecule of peroxide decomposed, the efficiency of polyethylene crosslinking should be accordingly lower than that of thermally initiated peroxide decomposition. The yield of free radicals should be moreover reduced by other side reactions. The amount of insoluble gel obtained in such a way does not exceed 40 %. (The efficiency of crosslinking reaction may, however, be, increased by a parallel use of polyfunctional monomers.)

The respective components of the redox-system may be applied separately. They may be allowed to react at a required stage of the polymer processing, e.g. after diffusion of transition metal ions into the polymer bulk. Hydroperoxide, polyfunctional monomer and other admixtures are usually mixed with the polymer before its shaping while an activator of crosslinking (a compound of transition metal) is deposited on the surface of the partially cooled product. By thermostating of the product in the absence of air, the activator penetrates into the polymer bulk and the decomposition reaction of hydroperoxide is accomplished.

Such a procedure of polyethylene crosslinking is suitable for thin products, as e.g. for foils of thickness 0.1 mm where the time of crosslinking of LDPE is about 1 h at 70 °C. To obtain more than 80 % of gel, 2.5 % w. of cumyl hydroperoxide, 2 % w. of glycol dimethacrylate and 0.1 % w. of Co(II) naphthenate should be used.

7.5 Microwave Induction

Initiation of crosslinking by microwave energy originates from the well-tried vulcanization of rubber mixtures by alternating electromagnetic fields of ultra-high frequencies. The capability of absorbing microwave energy is given by the permittivity and the loss factor of a substrate and these are determined by the polymer polarity. By absorption of microwave energy, material is warmed up and if a thermal initiator of free radical reactions is present in the system, crosslinking has the same character as a thermally initiated process. Since a polyolefin is nonpolar, it cannot be crosslinked by microwave energy without an addition of polar admixtures [109]. In many applications of crosslinked polyethylene, its nonpolar character plays a decisive role and polar admixtures should be avoided. A compromise involves the use of polar peroxides such as 4-chloro *tert.*butyl perbenzoate, di*tert.*butyl peroxy terephthalate, 3-*tert.*butyl peroxy-3 (4-chlorphenyl) phthalide [94, 110], etc. Crosslinking of polyethylene occurs at lower temperatures than would be expected from the decomposition parameters of these peroxides [26]. The decomposition is accelerated by the internal heating of peroxide molecules induced by increased dipole motions caused by the electromagnetic field of ultra-high frequency. Crosslinking may thus be accomplished faster than during the conventional heating of the system [111, 112].

7.6 Maleic Anhydride as Coagent

The crosslinking of polyethylene resulting from its reaction with peroxides is more profound if maleic anhydride is present in the system [113]. The crosslinked fraction contains a higher proportion of carbonyl groups from grafted maleic anhydride than uncrosslinked sol [114]. The amount of a gel will thus be affected not only by the number of crosslinks formed but also by a slower solubility of modified polyethylene in xylene which is brought about by polar groups grafted. This complies with the results of binding maleic anhydride to eicosane initiated by peroxides [115]. The grafts include single succinic rings. A high proportion of the product contains more than one succinic anhydride graft per eicosane molecule, suggesting thus that intramolecular reactions are important. The ratio of maleic anhydride units reacting with decomposing peroxide molecules is ten or more, so that propagation steps take place before termination can occur. It has been proposed that intermediates (excimers) formed in homopolymerization of maleic anhydride are responsible for the increased generation of polyethylene macroradicals and thus for an increased extent of the crosslinking reaction [116]. The reduction in the yield of a gel in the reaction of maleic anhydride with polyethylene in the presence of electron-donating compounds (amides, lactams, phosphites, sulphides, etc.) is assumed to be evidence that cationic species are involved in the reaction mechanism [117].

During the functionalization reaction of ethylene-propylene copolymers with maleic anhydride crosslinking is accompanied by a degradation of the polymer [118, 119] whereas in the case of polypropylene only degradation of the modified polymer occurs [120].

7.7 Azo Initiators

Aliphatic azoesters are other sources of free radicals potentially suitable for crosslinking of polyolefins. Since the temperature of their decomposition is higher than that of most thermally stable dialkyl peroxides, azo initiators are particularly suitable for crosslinking of products from high molecular mass polyethylene which requires higher processing temperatures. Gaseous low molecular mass compounds such as N_2, CH_4, and CO are formed during the decomposition of these compounds together with ketone and alkyl acetate; azo initiators can thus be used for the production of expanded crosslinked polyethylene products [121].

The group of azoesters considered as initiators of polyethylene crosslinking includes commercially available 2,2' azobis (2-acetoxy propane) and 2,2' azobis (2-acetoxybutane). Primary homolysis of 2,2' azobis (2-acetoxypropane)

$$
\begin{array}{c}
CH_3-C=O \quad O=C-CH_3 \\
| \qquad\qquad | \\
O \qquad\qquad O \\
| \qquad\qquad | \\
(CH_3)_2-C-N=N-C-(CH_3)_2
\end{array}
\quad \rightarrow N_2 + 2(CH_3)_2\dot{C}OCOCH_3 \qquad (27)
$$

leads to two acetoxypropyl radicals which may abstract a hydrogen atoms from the polyethylene chain and produce two alkyl macroradicals. A number of 2-acetoxypropyl radicals undergo fragmentation to acetone and acetyl radicals and, further to CO and methyl radicals. Methyl radicals can also react with the hydrocarbon chain to yield methane and alkyl macroradicals. The rates and decomposition temperatures of the two azoesters are given in Table 4 [122].

It may be noted that a frequently used thermal initiator of free radical polymerization of vinyl monomers, 2,2' azobis (isobutyronitrile) is not suitable for crosslinking

Table 4. Temperatures corresponding to different half-lives of decomposition of 2,2′ azobis (2-acetoxypropane) (*P*) and 2,2′ azo bis (2-acetoxybutane) (*B*)

Azoester	temperature (in °C) for half-life			
	10 h	1 h	10 min	1 min
P	189	215	238	270
B	191	217	239	271

because of its low decomposition temperature and the formation of radicals which have a relatively low reactivity in transfer reactions. Low reactivity of radicals primarily formed in decomposition of other azo initiators is a reason of lower crosslinking efficiency.

7.8 Reaction Systems for Polypropylene

Crosslinking of polypropylene initiated by thermal decomposition of peroxides is a more complicated task than a similar process with polyethylene. The main difficulty consists of the nature of radical reactions in polypropylene in which the majority of tertiary alkyl radicals undergo disproportionation and β-scission. Recombination of alkyl macroradicals occurs only as a side reaction; the efficiency of crosslinking is therefore low. The gel point can be attained when using either a large initial amount of peroxide (4 to 8 % w.) or an additional coagent such as a polyfunctional monomer, which can react with *tert*.alkyl macroradicals. By this route, a nature of the radical site is changed and the proportion of side reactions is reduced at the expense of recombination reaction. Incorporated unsaturated groups of monomers which may participate in subsequent reactions of alkyl macroradicals can thus have a positive effect on crosslinking. In such a case, the formation of the polymer network may be promoted by reactions of primary alkyl macroradicals with reactive groups of polyfunctional monomers.

Comparing the behaviour of 11 different vinyl and allyl monomers, it was found that pentaerythritol tetraallyl ether is the most effective as a coagent of polypropylene crosslinking; it increases the efficiency of crosslinking by a factor of 20, approximately, provided that the concentration ratio of reactants is optimal [123]. It is important to stress that the resulting crosslinking efficiency of a certain coagent is influenced considerably by the structure of the peroxide used. This indicates that a complex counter-play of different reactivity factors of primary radicals and polypropylene as well as of polyfunctional monomer and secondary macroradicals is involved [124]. A complete quantification of individual reactions is impeded by the fact that any polymer is non-uniform in its structure and consists of regions differing in the arrangement of the macromolecules and consequently in density and in solubility of the low molecular mass compounds. If several reactants are dissolved in a polymer, they may be dispersed nonhomogeneously as microdomains which, moreover, will differ from one reactant to the other. For the same reason, the reactivity of a polymer throughout the whole bulk of the sample will not be uniform. A model study of the effect of these factors was performed with a reaction of peroxides and organic sulfides in atactic polypropylene [125].

From other unsaturated organic compounds investigated, the most effective is 1,4-benzoquinone. The amount of gel formed depends significantly on the concentrations of both the peroxide and crosslinking coagent [126]. Increasing concentration of the initiator leads to an increase of the gel content at any coagent concentration. The dependence of the gel content on the initial amount of benzoquinone shows a maximum. This maximum is shifted to a higher concentration of benzoquinone on increasing the initial amount of peroxide. The existence of this maximum can be understood when we consider the reaction of benzoquinone with radicals formed from the decomposition of peroxide. At a high quinone concentration, this reaction competes successfully with the reaction of initiator radicals with polypropylene, which leads to the decrease of the amount of macroradicals available for crosslinking.

Another crosslinking system uses peroxide and sulfur or a sulfur compound where the yield of insoluble gel is very sensitive to the ratio of peroxide/sulfur [127]. The optimum for the formation of insoluble gel is reached for the ratio of dicumyl peroxide/sulfur close to 1:1. The role of sulfur in polypropylene crosslinking is similar to that of a multifunctional monomer. Sulfur suppresses side reactions of polypropylene macroradicals, and reduces the number of polymer main chain scissions. Addition of a multifunctional monomer as a third component of the crosslinking system does not increase the gel content, rather it reduces it. This effect can be explained when considering the scavenging of sulfur from the reaction system by its uneffective reactions with multifunctional monomers.

During crosslinking of polypropylene, initiated by peroxides conducted in the presence of thiourea and its derivatives it has been shown that the efficiency of the crosslinking coagent decreases parallel to the decrease of the sulfur content in the coagent [128]. This is in accordance with the fact that only sulfur was found in polypropylene crosslinked in the presence of thiourea, the amount of nitrogen being below the sensitivity threshold of elementary analysis. An assumption can be made that the increase of crosslinking efficiency of polypropylene initiated by peroxides and thiourea is caused by elementary sulfur formed in reactions of primary radicals and a sulfur containing coagent. The lower the ratio of sulphur related to other atoms of coagent, the higher the amount of dicumyl peroxide needed for the elimination of sulfur from the same molar amount of sulfur containing coagent.

New potential possibilities in polypropylene crosslinking which have not been elaborated very much are those from the use of azobisdicarbonamide as a crosslinking and blowing agent. This was attempted at temperatures ranging from 200 to 260 °C [129].

8 Crosslinking Through Silanes

At present, crosslinking of polyethylene through silanes is used mainly for the production of cable insulation and warm water pipe ducts. It is assumed that in cable production, the technology of polyethylene crosslinking based on silanes, which has many indisputable advantages, will predominate other procedures of polyethylene crosslinking as early as in the beginning of the 1990s [3].

Crosslinking consists of a condensation of silane groups bound to a polymer chain [130, 131, 132]. In the presence of water, silane groups, $Si(OX)_3$ involving a silicon atom surrounded by three alkoxyls OX, undergo hydrolysis [133]

$$H\overset{|}{\underset{|}{C}}-Si(OX)_3 + H_2O \rightarrow H\overset{|}{\underset{|}{C}}-Si(OX)_2OH + XOH \tag{28}$$

and alcohol XOH and silanol groups are formed. The latter enter a self-condensation reaction with the formation of siloxane crosslinks

$$H\overset{|}{\underset{|}{C}}Si(OX)_2OH + HO(XO)_2Si\overset{|}{\underset{|}{C}}H \rightarrow H\overset{OX}{\underset{OX}{C}}Si-O-Si\overset{OX}{\underset{OX}{C}}H + H_2O \tag{29}$$

From crosslinking agents, vinyltrimetoxy silane is the best available but other silanes may also be used [134].

Hydrolysis is relatively fast and may be induced even with traces of water absorbed in polyethylene. This is why a number of silane groups are already converted to silanols during polyolefin processing. Because of the low hydrophility of polyethylene, the extent of such a premature process is usually low [135]. A subsequent condensation of silanol groups is essentially slower than hydrolysis; it is even autoretarded when the first hydroxyls of silanol groups have reacted [136]. Under technological conditions, crosslinking therefore should be promoted by catalysts such as mineral and fatty acids, organic bases, esters of orthotitanic acid and some compounds of metals. The mechanism of formation of siloxane bridges which depends on the kind of the catalyst can involve three possible reaction pathways [137].

1. acid (HA)-catalyzed

$$-\overset{|}{\underset{|}{Si}}OH + HA \rightarrow -\overset{|}{\underset{|}{Si}}\overset{+}{O}H_2 + A^- \tag{30}$$

$$-\overset{|}{\underset{|}{Si}}OH + H_2^{\oplus}O\overset{|}{\underset{|}{Si}} + A^- \rightarrow -\overset{|}{\underset{|}{Si}}-O-\overset{|}{\underset{|}{Si}}- + H_2O + HA \tag{31}$$

2. base (B)-catalyzed

$$-\overset{|}{\underset{|}{Si}}OH + B \rightarrow -\overset{|}{\underset{|}{Si}}O^{\ominus} + \overset{+}{B}H \tag{32}$$

$$-\overset{|}{\underset{|}{Si}}OH + {}^{\ominus}O\overset{|}{\underset{|}{Si}}- + \overset{\oplus}{B}H \rightarrow -\overset{|}{\underset{|}{Si}}-O-\overset{|}{\underset{|}{Si}}- + B + H_2O \tag{33}$$

3. catalysis by neutral salts (AB)

$$2-\overset{|}{\underset{|}{Si}}OH \rightleftharpoons \overset{|}{\underset{|}{Si}} \begin{matrix} O-H \\ \diagup \cdots \cdots \\ A \\ \cdots \cdots \\ OH \leftarrow B \\ \diagup \\ -\overset{|}{Si}- \end{matrix} \rightarrow -\overset{|}{\underset{|}{Si}}-O-\overset{|}{\underset{|}{Si}}- + AB \cdot H_2O \qquad (34)$$

The original molecules of the catalysts are regenerated and condensation crosslinking steps may be repeated several times which affords satisfactory rates even at low catalyst content. From various potential catalysts of condensation crosslinking, organic compounds of tin and particularly dibutyl tin dilaurate, dibutyl tin dioctoate and dibutyl tin diacetate are used [138] since they are neutral, display a high catalytic activity and do not cause corrosion of the processing equipments. Besides the type and concentration of a catalyst, the rate and efficiency of crosslinking are affected also by the concentration of silane groups attached to the polymer, the distribution of the silane groups along the polymer chain, as well as the content of water and temperature of the reactor. Since higher rates of the formation of siloxane crosslinks have been observed with shorter alkoxy groups, methoxy derivatives of silanes are preferably used.

8.1 Functionalization of Polyolefins

Incorporation of silane groups into the hydrocarbon polymer can be performed in two ways, namely through a copolymerization of ethylene with vinyl trialkoxy silane [139]

$$-\dot{C}H_2 + CH_2=CHSi(OX)_3 \rightarrow -CH_2CH_2\dot{C}HSi(OX)_3 \qquad (35)$$

or as a grafting of vinyl trialkoxy silane to polymer. The latter method is older and more frequently requires the generation of macroradicals on the polyethylene chain in a similar way to peroxide-induced crosslinking. The only difference is that the process occurs in the presence of vinyl trialkoxy silane. In this case, macroradicals of polyethylene do not decay in recombination but add to a vinyl bond of silane [140]

$$\begin{matrix} | \\ CH_2 \\ | \\ \dot{C}H \\ | \\ CH_2 \\ | \end{matrix} + CH_2=CHSi(OX)_3 \rightarrow \begin{matrix} | \\ CH_2 \\ | \\ CHCH_2\dot{C}HSi(OX)_3 \\ | \\ CH_2 \\ | \end{matrix} \qquad (36)$$

and the resulting polymer contains side silane branches. A propagating radical of each branch decays in the transfer reaction with an other polymer segment and a new radical on a polymer chain, capable of entering polyaddition reaction with silane, is formed. The conditions are kept so that one initiating site can induce about 20–30 steps of the grafting reaction. A number of alkyl radicals, however, still recombine which causes an undesirable increase of viscosity of the polymer melt and a formation of microgels.

The choice of peroxide used is determined by the temperature of its decomposition. Peroxide should be effectively dispersed in the polymer melt before a substantial homolysis of O—O bonds can occur. For such a purpose, dicumyl peroxide which may be dissolved in vinyl trimethoxy silane (b.p. 120 °C) is suitable. The required degree of crosslinking was attained if 2% w. of silane with 5–10% w. of peroxide were added to polyethylene. At the silylation, grafting should not commence before both compounds (peroxide, silane) are well dispersed in the polymer melt [141]. Nonhomogeneous dispersion of additives reduces efficiency of grafting and of subsequent crosslinking.

8.2 Procedure of Crosslinking

The organometalic catalyst of crosslinking is added to the polymer before its processing usually as a master batch which contains also the necessary amount of antioxidant or metal deactivator. Other admixtures such as fillers may affect the crosslinking rather negatively. For example, hydrophilic fillers which reduce the amount of water in the polymer phase bring about the retardation of hydrolytic reactions since this reduction inhibits the formation of silanol groups which are the prerequisites of crosslinking [142]. Because of air humidity, silylated polyethylene has a limited storage time.

The rate of crosslinking which is controlled by the rate of diffusion of water into polyethylene depends on the nature of a crosslinked product and the temperature of the reaction. According to Fick's law, the time necessary to attain a certain degree of crosslinking is proportional to the square of the diffusion layer thickness, and inversely proportional to the product of relative humidity and the square of the diffusion coefficient of water in the polymer medium. Because of the nonpolar character of polyolefins and their crystallinity, diffusion of water is very slow. At a lower temperature, satisfactory rates of crosslinking are therefore achieved only for thin wall products of low density polyethylene. For example a cable insulation of 2.5 mm thickness requires as long as 6 days at 23 °C and 55% of relative humidity [3]. Crosslinking is therefore performed at higher temperatures (from 70 to 100 °C) where the process is faster: at 100 °C and in hot water crosslinking is accomplished in 4 hours. Besides the higher rates, the increased temperature also increases the crosslinking efficiency due to a higher proportion of the amorphous phase and thus of silane branches accessible to water. The accelerating effect of a higher temperature is beneficial in crosslinking of thinner products where the residual deformation of a polymer layer of the wound cable caused by its heating during crosslinking is not substantial [143]. The temperature 100 °C is used for an insulation thickness up to 8 mm, 80 °C for thicker insulations. Crosslinking at 80 °C, however, takes a relatively longer time (about 120 h for an insulation thickness of 8 mm, 260 hours for 15 mm and 1000 h for 22 mm). These data hold for silylated polyethylene prepared through grafting with vinyl trimethoxy silane; for an analogous copolymerization product, the rate of crosslinking is from 3 to 4 times lower. Because of higher crystallinity, the rate of crosslinking of high density polyethylene is lower, the time of crosslinking being double [144] that of low density polyethylene.

As has been already shown, the most serious obstacle occurring in technology of crosslinking through silanes is the relatively slow rate of the reaction. Various experiments have been conducted aimed at accelerating the crosslinking. Some improvement is achieved by the addition of 10 to 20% w. of ethylene-propylene copolymer to polyethylene which promotes an increase of the diffusion rate and consequently of crosslinking by a factor from 3 to 5.

In spite of these drawbacks, the described method represents a prospective procedure for polyolefin crosslinking which may be applied not only to polyethylene and ethylene copolymers but with some modifications to polypropylene and to other polyolefins with a branched main chain.

9 Polyolefin Blends

A general trend to widen the spectrum of properties of available macromolecular compounds by blending different types of polymers also exists within the polyolefin group [145, 146].

Polyethylene and polypropylene are mutually incompatible and the blend prepared from the melt of a mixture of the two polymers is, to a certain extent, heterogeneous. The transition of the original crystal structure of polyethylene into a pseudohexagonal modification depends on the irradiation dose, the dispersion method, and the conditions of the orientation of the macromolecule in the sample. Pseudohexagonal modification exists from 5 to 10 K over the melting temperature of a parent crystalline structure of polyethylene.

Mechanical properties depend not only on the crosslinking method but also on the rate of cooling of the polymer mixture [147, 148]. Polyethylene acts here as an impact strength modifier. The more finely it is dispersed in the blend, the better are the resulting mechanical properties [149].

The effect of ionizing irradiation on the changes of molecular mass of individual components of the mixture may be expected. Polypropylene molecules undergo degradation preferentially whereas low density polyethylene crosslinks. Of significance is the fact that the stability of the mechanical properties of the blend IPP/LDPE towards radiation is higher on increasing the content of amorphous phase by a faster rate of cooling [150].

The most pronounced decrease in the melting temperature is observed with isotactic polypropylene, indicating that the crystalline perfection of polypropylene is more highly affected by ionizing irradiation than that of polyethylene. Obviously, the process includes not only the main chain scissions but also the change in tacticity of polypropylene macromolecules.

Post-irradiation oxidative effects lead to a considerable deterioration of the mechanical properties of polypropylene and of its blends with a low content of polyethylene. The elongation at the breaking point of the polymer falls sharply even at doses of lower than 100 kGy which are used for example during the sterilization of medical instruments. In blends consisting of more than 30% w. of low density polyethylene, the influence of post-irradiation oxidative effects is markedly diminished [151].

Crosslinking of PE/PP blend films was performed by chemical reactions of dicumyl peroxide and electron-beam irradiation by using ultra-high molar mass polyethylenes (6.10^6 g/mol) and polypropylenes (4.10^6 g/mol) [152]. As a result, the crosslinking occurred in the polyethylene regions while in polypropylene regions it could not be achieved under any circumstances. It was found that the high temperature resistance of the polymer system had been improved. Accordingly it turned out that the heat resistance of the polymer blend had become worse as the polypropylene content had increased.

The crosslinking of blends IPP/LDPE induced by dicumyl peroxide up to 4% w. of its content incorporates only polyethylene into the polymer network [153]. In crosslinking of the 1:1 blend induced by 4% w. of *tert.* butyl perbenzoate, the insoluble gel contained 13% w. of polypropylene. If 2% w. of pentaerythritol tetraallyl ether (PETA) was used, the portion of PP in the gel increased from 50 to 70% w. The coagent does not particularly affect the crosslinking efficiency of the peroxide in polyethylene, but increases it significantly in polypropylene. The lower crosslinking efficiency of PETA coagent in the blends compared with polypropylene alone is associated with a better solubility of the coagent in the polyethylene phase. The crosslinking of polypropylene part then occurs with the lower initial concentration of PETA as it corresponds to the initial amount of PETA added to the mixture.

The decrese of the melting and crystallization temperatures of both PP and PE phases with an increasing concentration of peroxide cannot be ascribed to the increasing amount of crosslinks only, but, particularly, to other reactions of macromolecules.

10 The Effect of Crosslinking on Properties

10.1 Mechanical Properties

The molecular mass of a polymer increases during crosslinking and this is directly interrelated with the change of properties of the polymer connected with the sample deformability. The crosslinking leads to an increase of viscosity of the polymer melt, increase of the tensile strength [154], improvement of creep properties [155, 156], increase of the resistance against environment stress cracking [157], etc.

With regard to the mechanical reaction of a polymer network to a stress applied, it is important that loose ends of macromolecules in a network structure are as short as possible and/or their concentration is low. As these ends mostly extend out of the lamellas of crystallites then, while crosslinking is taking place in an amorphous phase and with the simultaneous presence of crystallites, a network with small loose ends should be formed. The crosslink junctions stabilize the natural molecular network (entanglements and crystallites), and every chain in the system is potentially elastically operative and can contribute to the stress in a tensile experiment [33]. The stabilization effect of chemical crosslinks on entanglements and crystallites may be the direct cause of observed differences in the determination of the amount of chemical crosslinks from mechanical property measurements and sol-gel analysis of the crosslinked polymer.

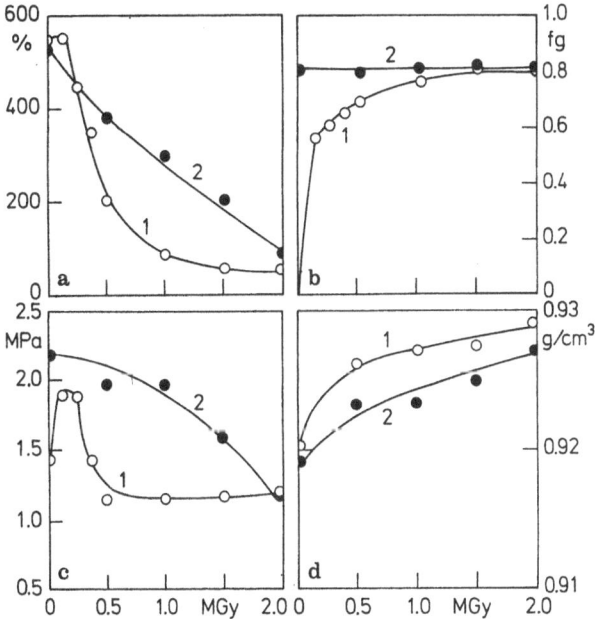

Fig. 5. Changes in **(a)** elongation at break (%), **(b)** gel fraction (fg), **(c)** tensile strength (MPa) and **(d)** density (gcm^{-3}) of (*1*) untreated and (*2*) pre-crosslinked (by 2% w. of dicumyl peroxide at 160 °C) LDPE as a function of gamma-irradiation dose

The concentration of crosslinks or the number n of the methylene CH$_2$ groups between two crosslink points is a universal molecular measure of the LDPE brittleness [30]. Elongation of polymer becomes less with a decrease of n: the significant reduction in the elongation at the break point occurs at n < 350. The plot of elongation on the number of crosslinks depends on the crosslinking method which conditions the distribution of crosslinks in a polymer sample (Fig. 5). A significant increase in elongation with the number of crosslinks has been observed with filled polymers which may be of importance when using highly filled polyalkenes in construction and building [158].

The formation of covalent bonds linking respective macromolecules leads simultaneously to an increase of the cohesion energy of a polymer system. A relative contribution of crosslinks to the cohesion energy becomes distinct, namely at temperatures above the melting temperature of the polymer crystallites. Amorphous regions of the polymer acquire higher structural strength by crosslinking and displacement of the chains in a crosslinked structure is more difficult. An external force applied to a crosslinked polyethylene product above the melting temperature of crystallites enforces a change in its shape, the product is then cooled below the crystallization temperature and the crystallites thus formed function as locked nuclei fixating new positions of macromolecules.

With the effect of a deformation force on a sample of crosslinked polymer molecule, only segments may be displaced irreversibly; the larger parts of the polymer system cannot be displaced without breaking the chemical bonds. For these reasons the shape memory and the permanency of the form is enhanced which means that an external deformation is fully reversible on releasing the force above the melting temperature of the crystallites. Macromolecular segments return to the thermodynamically most stable position after fusion of polymer crystallites.

Environmental stress cracking is seen on deformation of polymer samples in the presence of some solvents. Microcracks appear perpendicular to the applied stress. Even the number of 0.1 crosslinks per macromolecule reduces the probability of this phenomenon markedly; with increasing crosslink density this probability gradually decreases.

An important change in the properties of polyethylene which follows from the crosslinking is its increased resistance to creep which markedly prolongs the service-life of pressure pipes [159]. Crosslinking also positively affects the capacity of polyethylene to accept a higher amount of an inorganic filler without decreasing elongation at the breaking point.

10.2 Changes of Crystallinity

Crosslinking reduces the polyolefin crystallinity. Crosslinks play the role of defect centres which impede the folding of macromolecules and thus decrease the sizes of the lamellar crystals. Even a small number of crosslinks reduces the crystallinity of low density polyethylene provided that it is performed above the melting temperature of the polymer crystallites. The linkage of polymer chains impedes their independent displacement and the degree of crystallinity is distinctly lower even when only one macromolecule involves one crosslink [160].

An identical crosslink density does not necessarily lead to the same effect of crosslinking on properties. This is particularly evident in a case when crosslinks are formed between segments of polymer chains which are in the amorphous phase of semicrystalline polymer below the melting temperature of polymer crystallites. Here, the effect of crosslinking on crystallinity is lower.

It may be of interest that a small increase of polyethylene crystallinity has been observed for low degrees of crosslinking [161, 162]. This observation may be explained by a fixation of a given arrangement of macromolecules due to crosslinks formed. Scissions of macromolecules in the amorphous phase, where some physically entangled macromolecules may be released, which allows for an additional crystallization in a part of original amorphous phase may be another important factor contributing to this effect. A disruption of entangled chains is caused by mechanical strains among crystallites and by a transfer reaction of radicals to the strained polyethylene macromolecules.

The changes in crystallinity of polyethylene brought about by crosslinking are reflected in the melting temperature (T_m) [163, 164], heat capacity [165] and heat conductivity [166] of a polymer.

The reduction of the lamellar thickness of crystallites leads to a decrease of T_m. On the other hand, the crosslinks stabilize the arrangement of macromolecules which contributes to an increase of T_m. Actual values of T_m for a given sample of crosslinked polyethylene depend on initial crystallinity of a polymer, temperature at which the crosslinking is conducted, as well as on conditions of crystallization before a measurement of melting temperature of polymer crystallites. The effect of crosslinking on the heat capacity of polyethylene is mediated through the change of crystallinity and reduced mobility of macromolecules. The higher extent of amorphous parts of a polymer increases the heat capacity while the reduced mobility leads to its decrease. The resulting effect of crosslinking on the heat capacity depends on conditions of sample preparation and measurements. In the case of heat conductivity, the effect is similar. The crystalline phase has a higher heat conductivity than the amorphous phase. On the other hand, the formation of crosslinks brings about the higher density of packing of macromolecules which leads to an increase of the free pathway of phonons and the heat conductivity of crosslinked polyethylene therefore increases.

An initial increase and subsequent decrease of the diffusion coefficient of an antioxidant [167] and analogical picture obtained for sorption of solvents [168] in relation to the degree of crosslinking of LDPE was interpreted as being due two counter-acting effects, namely by a reduction in crystallinity and by an increase of packing of amorphous regions.

Besides the changes in crystallinity, the crosslinks change also the morphology of low density polyethylene [169]. The diameter of spherulites and the sheaf-shaped superstructure decreases with an increasing network. This result can be referred to as a decreased crystal growth rate related to an increase in the melt viscosity owing to an increasing network density.

10.3 Thermo-Oxidation Stability

A practical output of crosslinking consists in higher thermal stability of products at elevated temperatures without a deformation under a stress. Moreover, polyethylene with siloxane crosslinks has a temperature at which oxidation starts higher by more than 20 K [138].

The stabilization of samples of polyethylene to be crosslinked on irradiation brings some problems. Only a part of free radicals generated by irradiation contributes here to the formation of crosslinks in a fast reaction. The remaining macroradicals disappear not only by already mentioned side reactions but also in a diffusion controlled processes with oxygen and antioxidants. Carbonyl groups as characteristic products of the reaction of macroradicals with oxygen are formed not only on irradiating the polymer in air but also during irradiation in a nitrogen atmosphere [170]. The latter reaction may probably be ascribed to oxygen dissolved in a polymer. A relatively fast increase of concentration of carbonyl groups was observed over 1 year after irradiation crosslinking of polyethylene [171]. During oxidation initiated by macroradicals, peroxides and hydroperoxides are formed which are catalysts of thermal oxidation predominantly while carbonyl groups mediate photochemical oxidation. Oxidized chains are formed mainly on the surface of the irradiated polyethylene product, the thickness of an oxidized layer depending on irradiation conditions and on the type of polyethylene [172, 173]. The rate of carbonyl formation depends on the type of polyethylene: the greater the amorphous fraction the greater the formation of hydroperoxides and carbonyl groups. On the other hand, as estimated from mechanical properties, quenched samples show higher resistance to radiation. Hence, the stability of polyethylene depends not only on the density but also on the size of the crystallites [174]. The gel content of crosslinked polyethylene is unaffected by thermooxidative degradation during the induction period [175]: with a higher degree of oxidation the content of gel formed is considerably reduced [176].

Oxidation of polyethylene during irradiation manifests itself not only in deterioration of thermoxidative stability of the polymer and in ageing and fatigue of a material but the reaction of oxygen with macroradicals also leads to a decrease in the melting temperatures of LDPE and HDPE [177]. Melting behaviour of HDPE is changed by an appearance of a bimodal melting endotherm. The splitting of the melting endotherm is caused by the fact that in one part of the polymer bulk enriched by oxygenated products, the temperature of melting is depressed while in the part not attacked by oxygen, the melting temperature increases similarly as during irradiation of HDPE in vacuum. The different effect of irradiation on melting behaviour of HDPE and LDPE is caused mainly by slower diffusion of oxygen into HDPE.

10.4 Electrical Properties

Electrical and optical properties of polyalkenes are determined essentially by the structure of the main constituent parts of the polymer chain. The crosslinking which affects this structural level only a little therefore does not change these properties considerably. During the formation of crosslinks initiated by ionizing irradiation, an electric charge appears on the polymer. Radiation induced electric charging of a polymer is particularly effective with irradiation by electron beams. This fact should be borne in mind when using polymers [178] in radiation fields or during radiation modification of polymers.

During radical-initiated crosslinking various defects are created in the polymer and in its supermolecular structure by side reactions, which become nucleation centers for the formation of microvoids and are the reason for the failure of high voltage insulating properties [2]. Interestingly enough, water has been identified as a contributing factor (in the presence of voltage stress only), despite the fact that polyethylene is one of the most moisture resistant polymers available. Tree-like patterns occur on ageing, aptly called water (or electrochemical) trees.

10.5 Solubility and Swelling

Important properties which change significantly on crosslinking are the absorption of fluids and the solubility of crosslinked polyalkenes. The presence of crosslinks reduces the swelling capacity and hinders the transfer of polymer chains into a solution. The final degree of polymer swelling depends on expansivity of a macromolecular network, i.e. on the density of crosslinks [179]. A picture of the swelling process must take into account the fractal dimension of the network and the network strands in a dry state and in the presence of a solvent. With an increasing number of crosslinks the proportion of polymer sol decreases which is like the amount of a gel a measure of the crosslinking degree.

11 The Areas of Application of Different Crosslinking Procedures

The process of polyethylene crosslinking based on the thermal decomposition of peroxides, which is used in the production of high voltage cables, remains the most frequent one in industry [3, 180]. An increased thermomechanical resistance allows us to exploit the good insulating properties of the polymer even at high temperatures. This property becomes significant especially at higher operating and short-circuit loadings of power cables. The peroxidic method which is also used for crosslinking to a lower extent, e.g. for modification of the rheologic properties of a polyethylene melt, is used to produce highly expanded cellular materials. An increase in viscosity of the melt of a crosslinked polymer and a weaker dependence of viscosity on temperature enable us to produce highly expanded foams of polyethylene with a density of less than 20 kg m^{-3} in a continuous production process. For such a purpose, peroxide induced procedures, ionizing irradiation initiation [181], as well as silane crosslinking may be used, the last method being particularly useful for the preparation of polypropylene structural foams [182].

The principle of technological procedure during crosslinking induced by a thermal generation of free radicals consists of the thermal shaping of the product under conditions of relatively very slow decomposition of a crosslinking agent. Crosslinking is realized in a subsequent step on increasing the temperature of the system. The requirement to crosslink the polymer above the temperature of melting of polymer crystallites has some limitations. Not each shape of the final product can withstand such thermal treatment without deformation. For instance, during the production of pipes which, in contrast to cables, do not carry any support element, the crosslinking cannot be realized in the melt. Here, the crosslinking induced by ionization irradiation below melting temperature of polymer crystallites is preferred. For the same reason, this latter technology is also used for crosslinking of shrinkable products of a different shape; increasing density of crosslinks increases the shrinkage stress [183]. Crosslinking induced by betatrones is commercially used in the production of low voltage polyethylene cables.

For the production of thin shrinkable foils, photosensitized crosslinking may be used. The process may be performed with cheap equipment which does not need any particular safety precautions against radiation as it is, e.g. with sources of ionizing radiation. Photocrosslinked shrinkable foils are designed for insulation of gas conduits and oil pipelines [84].

Photocrosslinking is particularly suitable for coating other materials which cannot be exposed to high temperatures or for coating where heat is difficult to apply. As an example photocrosslinking of silicon-filled ethylene-propylene copolymers by α,α-dimethoxy phenyl acetophenone $C_6H_5COC(CH_3O)_2C_6H_5$ in air and in the presence of some polyfunctional monomers may be put forward [85].

The route via silanes is suitable for crosslinking of both the cables and pipes. It is becoming more and more preferred also in crosslinking of other products. Besides technological benefits, the silane method of crosslinking enables one to conduct crosslinking with such polyolefins where a network structure is difficult or even impossible to form with a radical mechanism.

Another research field which has not found a practical use up till now despite the large research effort is the crosslinking of oriented polyethylene macromolecules.

The preparation of polymeric fibres and films with high modulus has been extensively investigated and good results have been obtained for ultra high molecular weight polyethylene [184] (UHMWPE); the Young modulus of ultradrawn samples at 20 °C is nearly equal to the crystal lattice modulus of polyethylene. The temperature range of application of ultradrawn polyethylene fibers and films is limited by the low melting point of polyethylene. In conjunction with the improvement of thermal stability of UHMWPE, the crosslinking should also improve creep properties which are sometimes more important.

Crosslinked ultra high molecular-weight polyethylene fibers could be prepared by ionizing radiation without destroying the microstructure of the fibre. Unfortunately, radiation also induced the main chain scissions of oriented macromolecules thus reducing the tenacity of the filaments [185, 186]. More acceptable properties of the fibers may be obtained when ionizing radiation is applied to dried gel films of polyethylene preceding the fibers finishing stretch, the process being carried out in a nitrogen atmosphere [187, 188]. Porous gel-spun UHMWPE fibers consist of large folded chain lamellas linked to each other by means of tie macromolecules. During the irradiation the tie molecules are preferentially cleaved and new crosslinks between consecutive lamellas are formed [189].

Since a number of main chain scissions is lower when polyethylene is crosslinked by peroxides than on irradiating by ionizing radiation, dicumyl peroxide was used for crosslinking of UHMWPE fibers instead of irradiation [190]. Introduction of peroxide into the as-spun fibers was performed by swelling the fibers in a solution of dicumyl peroxide in hexane. Porous as-spun fibers were then prepared by spinning a solution of 5% (w/w) of polyethylene in a paraffin oil at 170 °C; the solvent was extracted from the fibers at room temperature by hexane. A relatively very high concentration of peroxide (up to 50% w. related to polyethylene) has been used for crosslinking; 1% of peroxide gave 16% of insoluble gel, 8% still yielded a relatively low value — 69% w. Very low efficiency of crosslinking indicates a nonuniform dispersion of peroxide in the polymer bulk. The decomposition of peroxide is likely to occur in macropores of fibers and in amorphous regions of polyethylene. Besides the intermolecular crosslink formation, the reaction leads to a relatively high proportion of cumyloxy groups bound to the polymer chain.

Crosslinked filaments display a nonfibrillating fracture surface and good resistance to high temperatures. It was found that crosslinked filaments can be kept at 195 °C for a prolonged period without significantly affecting the tenacity of the fiber at 20 °C.

According to recent reports [191], dicumyl peroxide (up to 200% w. in polymer) was introduced into UHMWPE gel in the swollen state of the polymer containing decalin. When the solvent is evaporated, dry gel films can be readily elongated to the desired draw ratio at 150 °C in a nitrogen atmosphere (and/or sulphuric acid). The crosslinking was carried out during elongation of gel films containing dicumyl peroxide. The specimen drawn up to 100 fold and crosslinked with 40% of dicumyl peroxide had a strength modulus greater than 110 GPa at 20 °C and still a value of 2 GPa even at 200 °C. At room temperature, this value is comparable with the Young modulus of copper and is much higher than that of aluminium and cast iron. Furthermore, the value at 200 °C is higher than the Young modulus of commercial high density polyethylene (1.3 GPa) at room temperature.

In a crosslinked sample, the hexagonal crystalline phase withstands even 230 °C. This apparently abnormally high temperature of melting may be explained by the fact that polymer chains were forced to retain an extended chain arrangement within the crystallites of the highly crosslinked and chemically modified amorphous domains and the entropy of their fusion would thus be smaller than the value for a random coil in the melt. In amorphous domains, a denser structure of crosslinks is probably formed and this has a high T_g temperature.

12 Concluding Remarks

Polyalkenes belong to the group of polymers which occupy one of the foremost places among thermoplastics in the volume of annual worldwide production. Interest in these polymers is due to their valuable properties and easy processability at relatively low production costs. Crosslinking as one of the post-modifications leading to an improvement in their useful properties further enlarges the application field of polyalkenes. The study of crosslinking reactions and adjacent changes of properties is highly popular among researchers. Analysis of the trends of common interest, however, indicates that there is a shift in partial research goals as well as in experimental methods. In the crosslinking of polyethylene we may thus say that the research effort devoted to the effect of gamma irradiation on crosslink formation and on changes of chemical structure related to the mechanical properties of irradiated polyethylene has grown tremendously in recent years. In spite of this, new knowledge which gives a more profound understanding of the chemistry of radiation crosslinking is still necessary.

Fundamental studies concerning peroxide induced crosslinking have been accomplished as well, this does not, however, exclude the need for new approaches. Chemical crosslinking need not necessarily be performed only with peroxides or thermally labile azo compounds as precursors of free radicals. In the study of the effect of chlorparaffins and antimony oxide on flammability of polyethylene was, e.g., demonstrated that polyethylene crosslinks to a high degree in the temperature range 250–350 °C [192]. Below and above this temperature range, the formation of a crosslinked network was not observed. The crosslinking can be induced by chlorparaffin itself; antimony oxide only enhances it. Chlorparaffins are incorporated into an insoluble network structure during this process. Thus it is evident that initiating systems for the crosslinking of polyolefins could be substantially more numerous and in the search for new crosslinking initiators more advantageous systems could be found.

Functionalization of polyalkenes oriented towards a subsequent crosslinking is another field insufficiently explored till now. Silylation of polyalkenes or copolymerization of cyclic dienes with olefins are examples of successful pathways. In the first procedure, the hydrolytic polycondensation of alkoxy silane groups is used for crosslinking while in the second case, the presence of cyclic diene facilitates crosslinking by sulfur. The potential possibilities, which are numerous, can be exemplified by the incorporation of cinnamic groups with subsequent photocrosslinking [193] in polyurethanes containing pendant quinoid groups [194] or on the condensation reaction of other incorporated reactive groups. From this large field of potentially available alternatives, the choice of new procedures will be determined by new requirements of techniques, the inventiveness of research workers, economical benefits, competition among manufacturers, and by many other factors.

Worth of note is also the relation between the character of the crosslinks and the polyolefin properties. A more detailed investigation of polyalkene networks containing polysulfidic bonds seems to be of use in this connection, since it was shown in the case of natural rubber [195, 196] that vulcanizates have a higher strength than peroxide or radiation crosslinked materials. This was ascribed to the lability of the sulfursulfur bond, which can presumably cleave and be reformed by bond interchange — thus absorbing the stress and preventing failure, a process which is probably not available to the carbon-carbon bond of peroxide or radiation crosslinking.

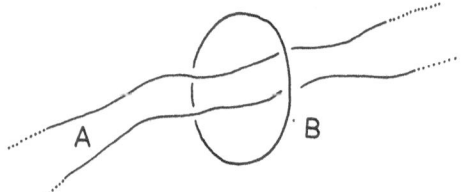

Fig. 6. Illustration of linking of macromolecules via cyclic oligomer: **A** — segments of a pair of macromolecules (dashed line represents the continuation of macromolecular chains), **B** — cyclic oligomer

It may be noted that besides chemical crosslink formation we can achieve the physical linking of macromolecules (Fig. 6) where the effect on properties could be just as important as in chemical crosslinking and entanglements. The pertinence of such a procedures may be demonstrated by the synthesis of polyrotaxanes [197] in which a cyclic oligomer of ethylene oxide is absorbed on macromolecules of polyamide (Scheme 1)

$$\left[\begin{array}{c} (CH_2CH_2O)_{12} \quad O \\ \\ NH{-}(CH_2)_{10}{-}C \end{array} \right]_n \qquad \text{Scheme 1}$$

Methods of cyclization of linear molecules may be vary. As regards physical linking of polyalkenes, the intramolecular end-to-end reaction of photoactive terminal groups linked by polymethylene chains could be of interest [198].

Such polymethylene compounds with dibenzaazepine chromophores and with n = 26 have quantum yields of cyclization 0.35 and of dimerization 0.1.

For the functionalization of polyalkenes with cycle forming structural units, other more complicated ways of catenane linking of macromolecules can be devised and it is only a question of time which of them will be used in the crosslinking of polymers.

13 References

1. Chapiro A (1988) Polymer Preprints, 29: 196
2. Bernstein BS (1988) Polymer Preprints, 29: 115
3. Cartasegna S (1986) Rubber Chem. Technol. 59: 722
4. Schnecko H (1979) Angew. Makromol. Chem. 76/77: 1
5. Hummel K (1979) Angew. Makromol. Chem. 76/77: 25
6. Dolgoplosk BA, Korshak YV (1984) Usp. Khim. 53: 65
7. Hummel K, Hubmann E, Pongratz T (1988) Europ. Polym. J. 24: 141
8. Baulch DL, Chouwn PK, Montangue DC (1979) Int. J. Chem. Kinet. 11: 1055
9. Gibian MJ, Corley RC (1973) Chem. Rev. 73: 441
10. Bakac A, Espenson JH (1986) J. Phys. Chem. 90: 325
11. Fujisaki N, Gauman T (1982) Int. J. Chem. Kinet. 14: 1059
12. Arthur NL, Christie JR (1987) Int. J. Chem. Kinet. 19: 261
13. Anastasi C, Arthur NL (1987) J. Chem. Soc., Farad. Trans. 2, 83: 277
14. Pritchard GD, Johnson KA, Nilsson WB (1985) Int. J. Chem. Kinet. 17: 327
15. Kopecky KR, Yeung MY (1988) Canad. J. Chem. 66: 374
16. Schuh H, Fischer H (1976) Int. J. Chem. Kinet. 8: 341
17. Bennet JE, Gale LH, Hayward EJ, Mile B (1973) J. Chem. Soc., Faraday Trans. 69: 1655
18. Tilman P, Tilquin B, Claes P (1982) J. Chim. Phys. 79: 629
19. Atarot H, Faucitano A, Cesca S (1976) Europ. Polym. J. 12: 169
20. Scholtens BJR (1984) J. Polym. Sci., Polym. Phys. Ed. 22: 317
21. Larina TG, Klinshpont ER (1988) Vysokomol. Soed. A 30: 1269
22. Šimunková D, Gál E (1988) Sulphur containing antioxidants in crosslinked polyethylene, 7th IUPAC Conference on Modified Polymers, Bratislava, 1988, p 19, p 37
23. Isakovich VN, Naumova SF (1988) Plast. Massy, (in russian), 7: 47
24. Šimunková D, Rado R, Šaliga A (1980) Plaste und Kautschuk, 27: 247
25. Lee DW, Braun D (1978) Angew. Makromol. Chem. 68: 199
26. Voight HU (1981) Kautschuk Gummi, Kunststoffe 34: 197
27. Voigt HU (1976) Kautschuk Gummi Kunststoffe, 29: 17
28. Polyethylene-Visico, Technical Information of Neste Chemicals, Stenungsund, 1987
29. Maillefer-Monosil, Technical Information of Maillefer, Ecublens, 1979
30. Nishimoto S, Kagyia VT (1986) Polym. Degrad. Stab. 15: 237
31. Vokál A (1986) Radiat. Phys. Chem. 28: 489
32. Zamotaev PV, Granchak VN, Litsov NI, Kachan AA (1985) Vysok. Soed. A 28: 2072 and references quoted on photocrosslinking
33. Klein PG, Ladizesky NH, Ward IM (1987) Polymer 28: 393
34. Miller DR, Macosko CW (1987) J. Polym. Sci., Polym. Phys. Ed. B 25: 2441
35. Miller DR, Macosko CW (1988) J. Polym. Sci., Polym. Phys. B 26: 1
36. Dole M (1981) J. Macrom. Sci., Chemm 15: 1403
37. Baird WG Jr, Joonase P, Rose AB, Helman WP (1982) Radiat. Phys. Chem. 19: 339
38. Chapiro A (1988) Nuclear Inst. Methods Phys. Res., B 32: 111
39. Ungar G (1981) J. Materials Sci. 16: 2635
40. Perez E, Vanderhart DL (1988) J. Polym. Sci. B 26: 1979
41. Klinshpont ER, Koryukhin VP, Milinchuk VK (1980) Vysokomol. Soed. A 22: 1754
42. Silverman J, Zoepfl FJ, Randall JC, Markovic V (1983) Radiat. Phys. Chem. 22: 583
43. Randall JC, Zoepfl JC, Silverman J (1983) Radiat. Phys. Chem. 22: 183
44. Okamoto H, Iwai T (1981) Radiat. Phys. Chem. 18: 407

45. Babic D (1988) Proceedings of 7th Bratislava IUPAC Conference on Modified Polymers, Bratislava 1988, also to be published in Makromol. Chem., Supplement C
46. Belyayev VM, Kogan SI, Pukshanskii MD, Budtov VP, Zemskova AP (1989) Vysok. Soed. A 31: 165
47. Hayakawa K, Kawase K, Yamakita H (1983) Radiat. Phys. Chem. 22: 929
48. Bartoš J, Tiňo J (1984) Polymer 25: 274; (1986) Polymer 27: 281
49. Klimová M, Szocs F (1988) Proceedings of the 7th Bratislava Conference on Modified Polymers, Bratislava, 1988, P 16, p 32
50. Basheer R, Dole M (1983) J. Polym. Sci., Polym. Phys. Ed. 21: 957
51. Pearson DS, Graessley WW (1980) Macromolecules 13: 1001
52. Basheer R, Dole M (1983) J. Polym. Sci., Polym. Phys. Edit. B 21: 949
53. Luo Y, Wang G, Lu Y, Chen N, Jing B (1985) Radiat. Phys. Chem. 25: 359
54. Vokál A (1988) Proceedings of the 7th Bratislava IUPAC Conference on Modified Polymers, Bratislava, 1988, P 21, p 39
55. Lacher LC, Bryant JL, Howard LN, Sumners DW (1986) Macromolecules 19: 2639
56. Sukhov FF, Feldman VI, Borzov SM, Slovokhotova NA (1988) Vysokom. Soed. A 30: 2213
57. Abramova IM, Kazalyan IM, Vatagina VA, Vasiliev VA, Tikhomirov VS (1986) Vysokom. Soed. A 28: 2019
58. Ahmad SR, Charlesby A (1978) Radiat. Phys. Chem. 11: 29
59. Ungar G (1980) Polymer 21: 1273
60. Mitsui H, Hosoi F, Kagyia T (1974) Polym. J. 6: 20
61. Leen DW, Braun D (1978) Angew. Makromol. Chem. 68: 199
62. Vasilienko VV, Klinshpont ER, Milinchuk VK (1978) Vysokom. Soed. A 20: 444
63. Vasilienko VV, Klinshpont ER, Milinchuk VK, Iskakov LI (1980) Vysokom. Soed. A 22: 1770
64. Melnikov MY (1982) J. Phys. Chem. (russian) 56: 2915
65. Shelukhov IP, Zhdanov GS, Klinshpont ER, Milinchuk VK (1988) Khim. Vysok. Energii (in russian) 22: 225
66. Ivanchenko VK, Margolin DM, Slovokhotova NA, Sukhor FF, Ilitcheva ZF, Terekhov VD, Leschenko SS (1985) Khim. Vysok. Energii (in russian), 19: 353
67. Basheer R, Dole M (1983) J. Polym. Sci., Polym. Phys. Ed. B 21: 949
68. Barron PF, Busfield WK, Hanna JV (1988) Polymer Comm. 29: 70
69. Briskman BA (1983) Uspekhi Khimii, (in russian), 52: 830
70. Kuroki T, Sawaguchi T, Niikumi S, Ikemura T (1982) Macromolecules 15: 1460
71. Hori Y, Kashiwabara H (1981) J. Polym. Sci., Polym. Phys. Ed. 19: 1141
72. Busfield WK, Appleby RW (1986) Brit. Polym. J. 18: 340
73. Chodák I, Zimányová E (1984) Europ. Polym. J. 20: 81
74. Klimová M, Tiňo J, Borsig E, Ambrovič (1985) J. Polym. Sci., Polym. Phys. Ed. 23: 105
75. Bartoš J, Tiňo J (1985) Coll. Czech. Chem. Commun. 50: 1391
76. Nojiri A, Sawasaki T (1985) Radiat. Phys. Chem. 26: 339
77. Busfield WK (1981) Europ. Polym. J. 17: 333
78. Sasaki Y, Imai M, Tanaka J, Shimazu H (1988) J. Polym. Sci. Chem. Ed. A 26: 2465
79. Okamura S (1981) Radiat. Phys. Chem. 18: 11
80. Wenxiu C, Goldman JP, Silverman J (1985) Radiat. Phys. Chem. 25: 317
81. Ranby B, Rabek JF (1977) ESR spectroscopy in polymer research Springer, Berlin Heidelberg New York, p 180
82. Ranby B, Hilborn J, Lie, CY, Jun QB (1988) Polym. Preprints 29: 526
83. Nehievich LA, Strelzova ZO (1983) Ukr. Khim. Zh. (in russian) 49: 650
84. Zamotaev PV (1988) Photoinitiated Crosslinking of Polyethylene, Proceedings of the 7th Bratislava Conference on Modified Polymers, Bratislava 1988, will also be published in Makromol. Chem., Supplement C
85. Hilborn J, Ranby B (1988) Rubber Chem. Techn. 61: 568
86. Zamotaev PV, Litzov NI, Kachan AA (1985) Khim. Vysok. Energii 19: 517
87. Zamotaev PV, Yakovlev VB, Litzov NI, Kachan AA, Komaschenko VN, Maystrenko AS (1983) Zh. Prikl. Spektroskopii (in russian) 39: 813
88. Zamotaev PV, Litzov NI, Kachan AA (1986) Polym. Photochem. 7: 139
89. Bartenev GM, Aliguliev RM (1982) Vysokomol. Soed. A 24: 1842
90. Zamotaev PV, Strelzova ZO, Kachan AA (1988) Plast. Massy (in russian) July: 37,

91. Schaller R, Martl MG, Hummel K (1987) Europ. Polym. J. 23: 259
92. Noury Initiators/delivery program, Technical Information of Akzo Chemicals, GmbH, Amersfoort 1987
93. LDPE Vernetzung mit Interox BCUP in Vergleich mit Interox DCUP. Technical report of Peroxid Chemie, Hollviegelskreuth, 1982
94. Dorn M (1982) Gummi Asbest Kunststoffe 35: 608
95. Myshkovskii VI, Nudelman ZN, Antonovskii VK (1980) Kautschuk i resina (in russian) 10: 20
96. Lazàr M (1983) Solid state reactions of peroxides. In: Patai S (ed) Chemistry of functional groups, peroxides, J. Wiley, Chichester, p 777
97. Graeber M (1979) Drahtwelt 30: 8
98. Aaltanen M (1978) Wire J. 11: 64
99. Smart G (1979) Wire Ind. 8: 551
100. Smart G (1979) Wire J. 12: 77
101. Agren L, Rauer K (1983) Haustechn. Rundschau 1: 18
102. Gupta JP, Stefton MV (1986) J. Appl. Polym. Sci. 31: 2195
103. Gedde UW (1986) Polymer 27: 269
104. van Drumpt JD, Osterwijk HHJ (1976) J. Polym. Sci., Polym. Chem. Edition, 14: 1485
105. Chodàk I, Lazàr M (1982) Angew. Makromol. Chem. 106: 153
106. Borsig E, Szocs F (1981) Polymer 22: 1400
107. Woods DW, Busfield WK, Ward IM (1985) Plast. Rubb. Process. Appl. 5: 157
108. Chodàk I, Rado R (1975) J. Polym. Sci., Symposium 53: 133
109. Menges G, Kircher K (1979) Kunststoffe 69: 430
110. Menges G, Strauch T (1985) Plastverarbeiter 36: 126
111. Microwave Curing Process of Polyethylene, Akzo Chemie, GmbH Düren 1983
112. Microwave Dry Curing, Brochure by Ing. P. Troester & Co. Maschinenfabrik, Hannover 1982
113. Gaylord N, Mehta M (1982) J. Polym. Sci., Polym. Lett. Edit. 20: 481
114. Gaylord N, Mehta M, Kumar V (1982) In: Corraher EJr (ed) Polymer science and technology 21: 171; Plenum, New York
115. Russell KE (1988) J. Polym. Sci. A 26: 2273
116. Gaylord N, Mehta R, Kumar V, Tazi (1988) Polymer Preprints 29: 565
117. Gaylord N, Mehta R (1988) J. Polym. Sci., A, Polym. Chem. 26: 1189
118. Gaylord N, Mehta M, Mehta R (1987) J. Appl. Polym. Sci. 33: 2549
119. De Vito G, Lanzetta N, Maglio G, Malinconico M, Musto P, Palumbo R (1984) J. Polym. Sci. Phys Chem. Ed. 22: 1335
120. Gaylord NG, Mishra MK (1983) J. Polym. Sci. Polym. Lett. Ed. 21: 23
121. Agren L, Rauer K (1983) Haustechn. Rundschau 1: 18
122. Luazo AP, Technical Information, Luperox GmbH, Günzburg, 1984
123. Borsig E, Fiedlerovà A, Lazàr M (1981) J. Macromol. Sci. Chem. A 16: 513
124. Capla M, Borsig E, Lazàr M (1987) Chem. papers 41: 253
125. Shlyapnikov VY, Kolesnikova NN (1987) Europ. Polym. J. 23: 633
126. Chodàk I, Lazàr M (1986) J. Appl. Polym. Sci. 32: 5431
127. Capla M, Borsig E, Lazàr M (1985) Angew. Makromol. Chem. 133: 53
128. Borsig E, Capla M, Lazàr M (submitted for publication) Polymer
129. Tuleshkov N, Dyumalusky S, Kotsev G (1988) Proceedings of Prague Microsymposium on Long Term Stability and Ageing of Polymers, Prague 1988, p 72
130. Weber BW (1978) Bull. SEV, 69: 62
131. Chvatova TP et al (1980) Crosslinking of polyethylene through organosilanes (in russian), Khimyia
132. Bloor R (1981) Plas. Technol. 27, p. 83 Jan. 83
133. Pluedemann EP (1982) Silane coupling agents, Plenum, New York
134. Hetflejš J (1983) Chem. Listy (in czech), 77: 843
135. Nishiyama N, Horib K (1987) J. Appl. Polym. Sci. 34: 1619
136. Smith KA (1987) Macromolecules 20: 2514
137. Konoval JV, Konovalenko NG, Ivanchev SS (1988) Uspekhi Khimii (in russian) 57: 134
138. Vasilec LG, Lebedeva ED, Akutin MS, Kazurin VJ (1988) Plast. Massy, July, 43
139. Fisher EJ (1987) Si-Link technology. Wire and Cable Seminar UCC, Oct., Budapest
140. Scott HG, Humphries JF (1973) Mod. Plastics 3: 133

141. Weber BW (1978) Bull. SEV 69: 62
142. Rozen MR (1978) J. Coatings Technology 50: 70
143. Sioplas, Tech. Information AEI Cables Ltd., Gravesend 1987
144. Dow Corning-Sioplas Technology, Tech. Information 1977
145. Spadaro G, Rizzo G, Acierno D, Caldero E (1984) Radiat. Phys. Chem. 23: 445
146. Antipov EM, Kupcov SA, Kuzmin NN, Pavlov SA (1988) Vysok. Soed. A 30: 1448
147. Galeski A, Bartzak Z, Pracella M (1984) Polymer 25: 1323
148. Teh JW (1983) J. Appl. Polym. Sci. 28: 605
149. Wenig W, Scholler T (1985) Progr. Colloid Polym. Sci. 71: 113
150. Kostoski D, Babic D, Stojanovic Z, Gal O (1986) Radiat. Phys. Chem. 28: 269
151. Kostoski D, Katcharevich-Popovich Z (1988) Polymer Commun. 29: 142
152. Sawatari C, Matsuo M (1987) Polym. J. 19: 1365
153. Borsig E, Fiedlerová A, Rychlá L, Lazár M, Rätzsch M, Haudel G (1989) J. Appl. Polym. Sci. 37: 467
154. Andreopoules AG, Kampouris EM (1986) J. Appl. Polym. Sci 31: 1061
155. Appleby RW, Busfield WK (1986) Polym. Commun. 27: 45
156. Bhateja SK, Andrews EH (1987) J. Appl. Polym. Sci. 34: 2809
157. Efimov AV, Valiotti NN, Dakin VJ, Ozerin AN, Bakeev NF (1988) Vysokomol. Soed. A 30: 963; Vysokomol. Soed. (1988) A 30: 2165
158. Knizhnik EI, Gordyenko VP, Ilyenko RE, Onisko AD (1983) Radiat. Phys. Chem. 22: 645
159. Poschet G (1987) Kunststoffe 77: 792
160. Bolder G, Meier M (1986) Kautschuk Gummi Kunststoffe 39: 715
161. Abramova JM, Kazaryan LG, Vatagina VA, Vasiliev VA, Tikhomirov VS (1986) Vysok. Soed. A 28: 2019
162. Stoyanov OV, Deberdeyev RY (1987) Vysokomol. Soed. B 29: 22
163. Zoepel FJ, Markovich V, Silverman J (1984) J. Polym. Sci., Polym. Chem. Ed. 22: 2017
164. Narkis M, Raiter I, Shkolnik S, Siegman A, Eyerer P (1987) J. Macromol. Sci., Phys. B 26: 37
165. Ribes-Greus A, Diaz-Calleja (1987) J. Appl. Polym. Sci. 34: 2819
166. Brikman BA (1983) Uspekhi Khimii (in russian) 52: 830
167. Shew K, Johnson JF (1988) Polym. Preprints 29: 117
168. Stoyanov OV, Polyanskii AA, Deberdeyev RY, Chalykh AE (1985) Vysokomol. Soed. 27 A: 1977
169. Bongardt J, Michler GH, Naumann I, Schulze G (1987) Angew. Makromol. Chem. 153: 55
170. Vacek K, Vokál A (1988) Free Radicals in Radiation Crosslinked Polyethylene Insulation Blends, Proceedings of 7th Bratislava IUPAC Conference on Modified Polymers, Bratislava July 1988, P 20, p 38
171. Birkinshaw C, Buggy M, Daly S (1988) Polym. Degrad, Stab. 22: 285
172. Audoin-Jiráčková L, Papet G, Verdu J (1989) Europ. Polym. J. 25: 181
173. Belousova MV, Maklarov AI, Skirda VD, Bykov EV, Romanov BS (1986) Vysokomol. Soed. 28 A: 663
174. Geetha R, Torikai A, Yoshida S, Nagaya S, Shirakawa H, Fueki K (1988) Polym. Degrad. Stab. 23: 91
175. Kramer E, Steiner G, Koppelman J (1988) Proceedings of the IUPAC Microsymposium on Chemical and Physical Phenomena in the Ageing of Polymers, SL 3, Prague, July 1988
176. Hikmet R, Keller A (1987) Radiat. Phys. Chem. 29: 15
177. Zoepfl FJ, Markovich V, Silverman J (1984) J. Polym. Sci., Polym. Chem. Ed. 22: 2033
178. Prokopyev OV, Rozno AG, Gromov VV (1988) Plast. Massy, 5: 19
179. Schmit C, Cohen-Addad JP (1989) Macromolecules 22: 142
180. Kertscher E (1981) Kunststoffe 71: 677
181. Sagane N, Harayama H (1981) Radiat. Phys. Chem. 18: 98
182. Lee YD, Wang LF (1986) J. Appl. Polym. Sci. 32: 4639
183. Mascia L (1987) Electron beaming of polymers for advanced technologies. In: Martuscelli E, Marchetta C, Nicolais L (eds) Future trends in polymer science and technology, Technomic Publ., Lancaster, p 25
184. Matsuo M, Sawatari C (1986) Macromolecules 19: 2036
185. de Boer J, Pennings AJ (1981) Polym. Bull. 5: 317; (1982) Polym. Bull. 7: 309
186. Sawatari C, Matsuo M (1985) Colloid Polym. Sci. 263: 783

187. Sawatari C, Nishikido H, Matsuo M (1988) Colloid Polym. Sci. 266: 316
188. Lemstra PJ (1988) Proceedings of the 7th IUPAC Bratislava International Conference on Modified Polymers, p. 105, Bratislava, July 1988
189. Dijkstra DJ, Pennings AJ (1988) Polym. Bull. 20: 557
190. de Boer J, van den Berg HJ, Pennings AJ (1984) Polymer 25: 513
191. Matsuo M, Sawatari C (1986) Macromolecules 19: 2028; (1987) Macromolecules 20: 1745
192. Fedeev SS, Morozova LS, Drobinin AN, Bogdanova VV, Klimovcova IA, Lesnikovich AI (1988) Vysokomol. Soed. A 30: 2180
193. Lemaitre E, Coqueret X, Mercier R, Lablache-Combier A, Loucheux C (1987) J. Appl. Polym. Sci. 33: 2189
194. Cha Y, Tsunooka M, Tanaka M (1986) J. Photochem. 35: 93
195. Kempermann T (1988) Rubber Chem. Technol. 61: 422
196. Nasir M, Teh GK (1988) Europ. Polym. J. 24: 733
197. Gibson HW, Engen P, Lecavalier P (1988) Polym. Preprints p. 29
198. Ashikaga K, Ito S, Yamamoto M, Nishijima Y (1988) J. Amer. Chem. Soc. 110: 198

Editor: H.-J. Cantow
Received June 19, 1989

Author Index Volumes 1–95

Subject Index